ACKNOWLEDGEMENT

First and foremost, praises and thanks to Our Parents and Our Family Members, the God, the Almighty, and Our well-wishers for their showers of blessings throughout Our book work and to complete the publication successfully.

The completion of this undertaking would not have been possible without the participation and assistance of many people whose names may not be enumerated here. Their contributions are sincerely appreciated and gratefully acknowledged. However, the group likes to express their deep appreciation and indebtedness particularly to the following:

We pleased to express our deepest sense of gratitude to Dr.S.Rajalakshmi Correspondent, Dr. S.N.Subbramanian Chairman, Dr.S.Nalin Vimalkumar Technical Director, Dr.V.P.Arunachalam Director SNS Technical institutions, Coimbatore, Dr.S.Chenthur Pandian Principal ,SNS College of Technology for providing us a chance to bring out this book and good wishes .

We render our sincere thanks to family members, relatives, friends and others who were very much supportive both morally and physically during this publication. Above all, We render our heartfelt gratitude and thanks to the Great Almighty, the author of knowledge and wisdom, for his abundant love.

ARTIFICIAL INTELLIGENCE

(Common to All Engineering Branches)

As Per The Latest Syllabus of Anna University, Chennai (Regulation 2017)

Dr. L. M. Nithya

Professor & Dean (CSE&IT),
SNS College of Technology, Coimbatore,
Tamil Nadu, India.

Dr. M. Udhayamoorthi

Associate Professor,
Dept. of Information Technology,
SNS College of Technology, Coimbatore,
Tamil Nadu, India.

Mr. M. Jagadesh

Assistant Professor,
Dept. of Electronics & Communication Engineering,
SNS College of Technology, Coimbatore,
Tamil Nadu, India.

Mr. D. Vikneshkumar

Assistant Professor,
Dept. of Information Technology,
SNS College of Technology, Coimbatore,
Tamil Nadu, India.

SHANLAX
PUBLICATIONS

Title: Artificial Intelligence

Author's Name: Dr. L. M. Nithya,
Dr. M. Udhayamoorthi,
Mr. M. Jagadesh &
Mr. D. Vikneshkumar

Published by: Mr.S.Lakshmanan, Shanlax Publications, Vasantha Nagar, Madurai - 625003, Tamil Nadu India

Publisher's Address: 61, 66 T.P.K. Main Road, Vasantha Nagar, Madurai - 625003, Tamil Nadu, India

Printer's Details: Shanlax Press, 66 T.P.K. Main Road, Vasantha Nagar, Madurai - 625003, Tamil Nadu, India

Edition Details (I,II,III): I

ISBN: 978-93-91373-33-7

Pages: 256

Price: 320/-

CONTENTS

Chapter 1
INTRODUCTION

1.1 Introduction to AI

1.1.1 What is artificial intelligence?

Artificial Intelligence is the branch of computer science concerned with making computers behave like humans.

Major AI textbooks define artificial intelligence as "the study and design of intelligent agents," where an intelligent agent is a system that perceives its environment and takes actions which maximize its chances of success. John McCarthy, who coined the term in 1956, defines it as "the science and engineering of making intelligent machines, especially intelligent computer programs."

The definitions of AI according to some text books are categorized into four approaches and are summarized in the table below:

Systems that think like humans –The exciting new effort to make computers think ... machines with minds, in the full and literal sense.‖(Haugeland,1985)	**Systems that think rationally** –The study of mental faculties through the use of computer models.‖ (Charniak and McDermont,1985)
Systems that act like humans The art of creating machines that perform functions that require intelligence when performed by people.‖(Kurzweil,1990)	**Systems that act rationally** –Computational intelligence is the study of the design of intelligent agents.‖(Poole et al.,1998)

The four approaches in more detail are as follows:

a. Acting humanly: The Turing Test approach

- Test proposed by Alan Turing in 1950
- The computer is asked questions by a human interrogator.
- The computer passes the test if a human interrogator, after posing some written questions, cannot tell whether the written responses come from a person or not. Programming a computer to pass, the computer need to possess the following capabilities:

1. Natural language processing to enable it to communicate successfully in English.
2. Knowledge representation to store what it knows or hears
3. Automated reasoning to use the stored information to answer questions and to draw new conclusions.
4. Machine learning to adapt to new circumstances and to detect and extrapolate patterns

To pass the complete Turing Test,the computer will need

- **Computer vision** to perceive the objects,and
- **Robotics** to manipulate objects and move about.

b. Thinking humanly: The cognitive modeling approach

We need to get inside actual working of the human mind:

- through introspection – trying to capture our own thoughts as they go by;
- through psychological experiments

Allen Newell and Herbert Simon, who developed GPS, the General Problem Solver‖ tried to trace the reasoning steps to traces of human subjects solving the same problems.

The interdisciplinary field of cognitive science brings together computer models from AI and experimental techniques from psychology to try to construct precise and testable theories of the workings of the human mind

c. Thinking rationally: The "laws of thought approach"

The Greek philosopher Aristotle was one of the first to attempt to codify –right thinking‖, that is irrefutable reasoning processes. His syllogism provided patterns for argument structures that always yielded correct conclusions when given correct premises—for example, –Socrates is a man; all men are mortal; therefore Socrates is mortal.‖.

These laws of thought were supposed to govern the operation of the mind; their study initiated a field called logic.

d. Acting rationally: The rational agent approach

An agent is something that acts. Computer agents are not mere programs, but they are expected to have the following attributes also:

- operating under autonomous control,
- perceiving their environment,
- persisting over a prolonged time period,
- adapting to change.

A **rational agent** is one that acts so as to achieve the best outcome.

Insertion inside a sequential file is not possible. It is similar to the scenario that it is not possible to insert a song in the middle of a previously recorded tape. For updation of records, a new record should be written on a different tape with changes when record is updated.

Efficiency of sequential file organization The cfficicncy of locating data in sequential organization depends on the type of query. The best case is the location of record in the beginning of the tape and worst case

1.1.2 The foundations of Artificial Intelligence

The various disciplines that contributed ideas, viewpoints, and techniques to AI are given below:

Philosophy (428 B.C. - present)

Aristotle (384-322 B.C.) was the first to formulate a precise set of laws governing the rational part of the mind. He developed an informal system of syllogisms for proper reasoning, which allowed one to generate conclusions mechanically, given initial premises.

	Computer	Human Brain
Computational units	1 CPU,108 gates	1011 neurons
Storage units	1010 bits RAM	1011 neurons
	1011 bits disk	1014 synapses
	10-9 sec	10-3 sec
Cycle time	1010 bits/sec	1014 bits/sec
Bandwidth	109	1014
Memory updates/sec		

Table 1.1 A crude comparison of the raw computational resources available to computers (circa 2003) and brain. The computer's numbers have increased by at least by a factor of 10 every few years. The brain's numbers have not changed for the last 10,000 years.

Brains and digital computers perform quite different tasks and have different properties. Table 1.1 shows that there are 10000 times more neurons in the typical human brain than there are gates in the CPU of a typical high-end computer. Moore's Law predicts that the CPU's gate count will equal the brain's neuron count around 2020.

Psycology (1879 – present)

The origin of scientific psychology are traced back to the wok if German physiologist Hermann von Helmholtz(1821-1894) and his student Wilhelm Wundt(1832 – 1920)

In 1879, Wundt opened the first laboratory of experimental psychology at the university of Leipzig. In US, the development of computer modeling led to the creation of the field of cognitive science. The field can be said to have started at the workshop in September 1956 at MIT.

Computer engineering (1940-present)

For artificial intelligence to succeed, we need two things: intelligence and an artifact. The computer has been the artifact of choice.

A1 also owes a debt to the software side of computer science, which has supplied the operating systems, programming languages, and tools needed to write modern programs

Control theory and Cybernetics (1948-present)

Ktesibios of Alexandria (c. 250 B.c.) built the first self-controlling machine: a water clock with a regulator that kept the flow of water running through it at a constant, predictable pace.

Modern control theory, especially the branch known as stochastic optimal control, has as its goal the design of systems that maximize an objective function over time.

Linguistics (1957-present)

Modem linguistics and AI, then, were "born" at about the same time, and grew up together, intersecting in a hybrid field called computational linguistics or natural language processing.

1.1.3 The History of Artificial Intelligence

The gestation of artificial intelligence (1943-1955)

There were a number of early examples of work that can be characterized as AI, but it was Alan Turing who first articulated a complete vision of A1 in his 1950 article "Computing Machinery and Intelligence." Therein, he introduced the Turing test, machine learning, genetic algorithms, and reinforcement learning.

The birth of artificial intelligence (1956)

McCarthy convinced Minsky, Claude Shannon, and Nathaniel Rochester to help him bring together U.S. researchers interested in automata theory, neural nets, and the study of intelligence. They organized a two-month workshop at Dartmouth in the summer of 1956.

Perhaps the longest-lasting thing to come out of the workshop was an agreement to adopt McCarthy's new name for the field: artificial intelligence.

Early enthusiasm, great expectations (1952-1969)

The early years of A1 were full of successes-in a limited way. General Problem Solver (GPS) was a computer program created in 1957 by Herbert Simon and Allen Newell to build a universal problem solver machine. The order in which the program considered subgoals and possible actions was similar to that in which humans approached the same problems. Thus, GPS was probably the first program to embody the "thinking humanly" approach.

At IBM, Nathaniel Rochester and his colleagues produced some of the first A1 programs. Herbert Gelernter (1959) constructed the Geometry Theorem Prover, which was able to prove theorems that many students of mathematics would find quite tricky.

Lisp was invented by John McCarthy in 1958 while he was at the Massachusetts Institute of Technology (MIT). In 1963, McCarthy started the AI lab at Stanford. Tom Evans's ANALOGY program (1968) solved geometric analogy problems that appear in IQ tests, such as the one in Figure 1.1.

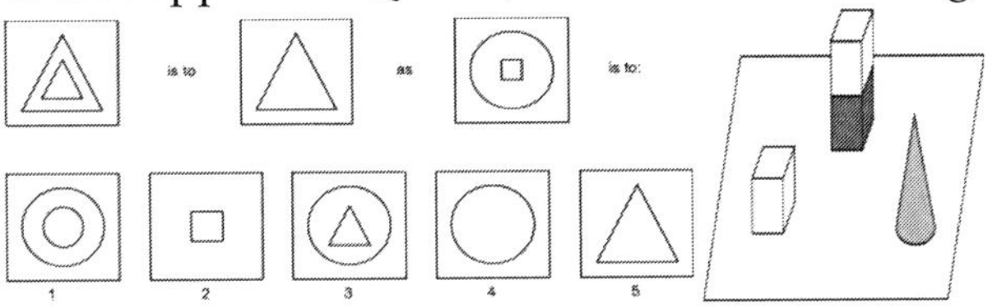

Figure 1.1 The Tom Evan's ANALOGY program could solve geometric analogy problems as shown

A dose of reality (1966-1973)

From the beginning, AI researchers were not shy about making predictions of their coming successes. The following statement by Herbert Simon in 1957 is often quoted:

It is not my aim to surprise or shock you-but the simplest way I can summarize is to say that there are now in the world machines that think, that learn and that create. Moreover, their ability to do these things is going to increase rapidly until-in a visible future-the range of problems they can handle will be coextensive with the range to which the human mind has been applied.

Knowledge-based systems: The key to power? (1969-1979)

Dendral was an influential pioneer project in artificial intelligence (AI) of the 1960s, and the computer software expert system that it produced. Its primary aim was to help organic chemists in identifying unknown organic molecules, by analyzing their mass spectra and using knowledge of chemistry. It was done at Stanford University by Edward Feigenbaum, Bruce Buchanan, Joshua Lederberg, and Carl Djerassi.

A1 becomes an industry (1980-present)

In 1981, the Japanese announced the "Fifth Generation" project, a 10-year plan to build intelligent computers running Prolog. Overall, the A1 industry boomed from a few million dollars in 1980 to billions of dollars in 1988.

The return of neural networks (1986-present)

Psychologists including David Rumelhart and Geoff Hinton continued the study of neural-net models of memory.

A1 becomes a science (1987-present)

In recent years, approaches based on hidden Markov models (HMMs) have come to dominate the area. Speech technology and the related field of handwritten character recognition are already making the transition to widespread industrial and consumer applications. The Bayesian network formalism was invented to allow efficient representation of, and rigorous reasoning with, uncertain knowledge.

The emergence of intelligent agents (1995-present)

One of the most important environments for intelligent agents is the Internet.

1.1.4 The state of art

What can A1 do today?

Autonomous planning and scheduling:

A hundred million miles from Earth, NASA's Remote Agent program became the first on-board autonomous planning program to control the scheduling of operations for a spacecraft (Jonsson et al., 2000). Remote Agent generated plans from high-level goals specified from the ground, and it monitored the operation of the spacecraft as the plans were executed-detecting, diagnosing, and recovering from problems as they occurred.

Game playing

IBM's Deep Blue became the first computer program to defeat the world champion in a chess match when it bested Garry Kasparov by a score of 3.5 to 2.5 in an exhibition match (Goodman and Keene, 1997).

Autonomous control

The ALVINN computer vision system was trained to steer a car to keep it following a lane. It was placed in CMU's NAVLAB computer-controlled minivan and used to navigate across the United States-for 2850 miles it was in control of steering the vehicle 98% of the time.

Diagnosis

Medical diagnosis programs based on probabilistic analysis have been able to perform at the level of an expert physician in several areas of medicine.

Logistics Planning

During the Persian Gulf crisis of 1991, U.S. forces deployed a Dynamic Analysis and Replanning Tool, DART (Cross and Walker, 1994), to do automated logistics planning and scheduling for transportation. This involved up to 50,000 vehicles, cargo, and people at a time, and had to account for starting points, destinations, routes, and conflict resolution among all parameters. The AI planning techniques allowed a plan to be generated in hours that would have taken weeks with older methods. The Defense Advanced Research Project Agency (DARPA) stated that this single application more than paid back DARPA's 30-year investment in AI.

Robotics

Many surgeons now use robot assistants in microsurgery. HipNav (DiGioia et al., 1996) is a system that uses computer vision techniques to create a three-dimensional model of a patient's internal anatomy and then uses robotic control to guide the insertion of hip replacement prosthesis. Language understanding and problem solving: PROVERB (Littman et al., 1999) is a computer program that solves crossword puzzles better than most humans, using constraints on possible word fillers, a large database of past puzzles, and a variety of information sources including dictionaries and online databases such as a list of movies and the actors that appear in them.

1.2 Intelligent Agents

1.2.1 Agents and environments

An agent is anything that can be viewed as perceiving its environment through sensors and SENSOR acting upon that environment through actuators. This simple idea is illustrated in Figure 1.2.

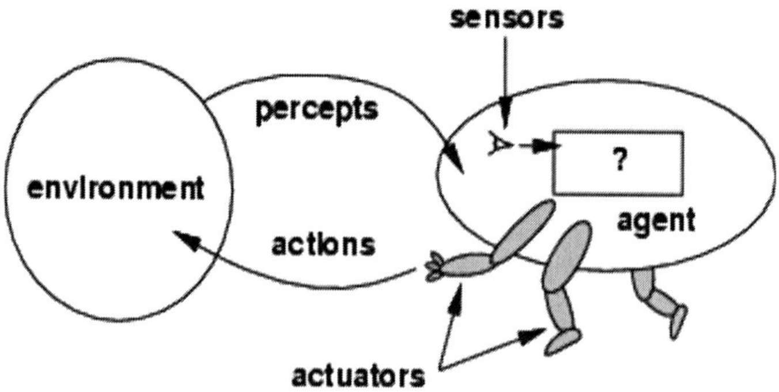

Figure 1.2 Agents interact with environments through sensors and actuators

- A human agent has eyes, ears, and other organs for sensors and hands, legs, mouth, and other body parts for actuators.

- A robotic agent might have cameras and infrared range finders for sensors and various motors for actuators.
- A software agent receives keystrokes, file contents, and network packets as sensory inputs and acts on the environment by displaying on the screen, writing files, and sending network packets.

Percept

We use the term percept to refer to the agent's perceptual inputs at any given instant.

Percept Sequence

An agent's percept sequence is the complete history of everything the agent has ever perceived.

Agent function

Mathematically speaking, we say that an agent's behavior is described by the agent function that maps any given percept sequence to an action. f: P*□A

Agent program

Internally, the agent function for an artificial agent will be implemented by an agent program. It is important to keep these two ideas distinct. The agent function is an abstract mathematical description; the agent program is a concrete implementation, running on the agent architecture.

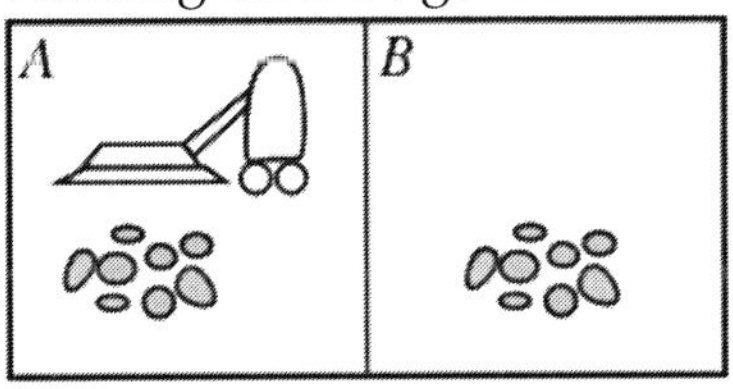

Figure 1.3 A vacuum-cleaner world with just two locations

To illustrate these ideas, we will use a very simple example-the vacuum-cleaner world shown in Figure 1.3.

Agent function

Percept Sequence	Action
[A, Clean]	Right
[A, Dirty]	Suck
[B, Clean]	Left
[B, Dirty]	Suck
[A, Clean], [A, Clean]	Right
[A, Clean], [A, Dirty]	Suck
...	

Figure 1.4 Partial tabulation of a simple agent function for the vacuum-cleaner world shown in Figure 1.3.

This particular world has just two locations: squares A and B. The vacuum agent perceives which square it is in and whether there is dirt in the square. It can choose to move left, move right, suck up the dirt, or do nothing. One very simple agent function is the following: if the current square is dirty, then suck, otherwise move to the other square. A partial tabulation of this agent function is shown in Figure 1.4.

Agent Program
Function REFLEX-VACUUM-AGENT ([Location, status]**)**
returns an action
if status = Dirty **then return** Suck
else if location = A **then return** Right
else if location = B **then return** Left

Rational Agent

A rational agent is one that does the right thing-conceptually speaking, every entry in the table for the agent function is filled out correctly. Obviously, doing the right thing is better than doing the wrong thing. The right action is the one that will cause the agent to be most successful.

Performance measures

A performance measure embodies the criterion for success of an agent's behavior. When an agent is plunked down in an environment, it generates a sequence of actions according to the percepts it receives. This sequence of actions causes the environment to go through a sequence of states. If the sequence is desirable, then the agent has performed well.

Rationality

What is rational at any given time depends on four things:

- The performance measure that defines the criterion of success.
- The agent's prior knowledge of the environment.
- The actions that the agent can perform.
- The agent's percept sequence to date.

This leads to a **definition of a rational agent** For each possible percept sequence, a rational agent should select an action that is expected to maximize its performance measure, given the evidence provided by the percept sequence and whatever built-in knowledge the agent has Omniscience, learning, and autonomy.

Omniscience, learning, and autonomy

An omniscient agent knows the actual outcome of its actions and can act accordingly; but omniscience is impossible in reality. Doing actions in order to modify future percepts-sometimes called information gathering-is an important part of rationality.

Our definition requires a rational agent not only to gather information, but also to learn as much as possible from what it perceives. To the extent that an agent relies on the prior knowledge of its designer rather than on its own percepts, we say that the agent lacks autonomy. A rational agent should be autonomous-it should learn what it can to compensate for partial or incorrect prior knowledge.

Task environments

We must think about task environments, which are essentially the "problems" to which rational agents are the "solutions."

Specifying the task environment

The rationality of the simple vacuum-cleaner agent, needs specification of

- the performance measure
- the environment
- the agent's actuators and sensors.

PEAS

All these are grouped together under the heading of the task environment. We call this the PEAS (Performance, Environment, Actuators, Sensors) description. In designing an agent, the first step must always be to specify the task environment as fully as possible.

formance Measure	Environments	Actuators	Se
gal, table trip, ize profits	Roads, other traffic, pedestrians, customers	Steering, accelerator, brake, Signal, horn, display	Cameras, sona GPS, Odomete sensors, keybo accelerometer

gure 1.5 PEAS description of the task environment for an automated taxi.

Performance Measure	Environments	Actuators	
Healthy patient, minimize costs, lawsuits	Patient, hospital, staff	Display questions, tests, diagnoses. Treatments, referrals	Key sym pati
Correct image categorization	Downlink from orbiting satellite	Display categorizatio n of scene	Col
Percentage of arts in correct bins	Conveyor belt with parts: bins	Joined arm and hand	Can sens
Maximize purity, yield, safety	Refine operators	Values, pumps, heaters, displays	Tem Pre
Maximum student's score on test	Set of students, testing agency	Display exercises, suggestions, corrections	Key

Figure 1.6 Examples of agent types and their PEAS descriptions

Properties of task environments

- Fully observable vs. partially observable
- Deterministic vs. stochastic
- Episodic vs. sequential
- Static vs. dynamic
- Discrete vs. continuous
- Single agent vs. multi agent

Fully observable vs. partially observable

If an agent's sensors give it access to the complete state of the environment at each point in time, then we say that the task environment is fully observable. A task environment is effectively fully observable if the sensors detect all aspects that are relevant to the choice of action;

An environment might be partially observable because of noisy and inaccurate sensors or because parts of the state are simplly missing from the sensor data.

Deterministic vs. stochastic

If the next state of the environment is completely determined by the current state and the action executed by the agent, then we say the environment is deterministic; otherwise, it is stochastic.

Episodic vs. sequential

In an episodic task environment, the agent's experience is divided into atomic episodes. Each episode consists of the agent perceiving and then performing a single action. Crucially, the next episode does not depend on the actions taken in previous episodes.

For example, an agent that has to spot defective parts on an assembly line bases each decision on the current part, regardless of previous decisions;

In sequential environments, on the other hand, the current decision could affect all future decisions. Chess and taxi driving are sequential:

Discrete vs. continuous

The discrete/continuous distinction can be applied to the state of the environment, to the way time is handled, and to the percepts and actions of the agent. For example, a discrete- state environment such as a chess game has a finite number of distinct states.

Chess also has a discrete set of percepts and actions. Taxi driving is a continuous state and continuous-time problem: the speed and location of the taxi and of the other vehicles sweep through a range of continuous values and do so smoothly over time.

Taxi-driving actions are also continuous (steering angles, etc.). *Artificial Intelligence*

Single agent vs. multi agent

An agent solving a crossword puzzle by itself is clearly in a single agent environment, whereas an agent playing chess is in a two-agent environment.

As one might expect, the hardest case is partially observable, stochastic, sequential, dynamic, continuous, and multi agent.

t	Observe	Deterministic	Episodic	Static	
e	Fully	Deterministic	Sequential	Static	Dis
	Fully	Strategic	Sequential	Semi	Dis
on	Partially	Stochastic	Sequential	Static	Dis
	Fully	Stochastic	Sequential	Static	Dis
	Partially	Stochastic	Sequential	Dynamic	Cor
s	Partially	Stochastic	Sequential	Dynamic	Cor
	Fully	Deterministic	Episodic	Semi	Cor
t	Partially	Stochastic	Episodic	Dynamic	Cor
troller	Partially	Stochastic	Sequential	Dynamic	Cor
nglish	Partially	Stochastic	Sequential	Dynamic	Dis

Figure 1.7 Examples of task environments and their characteristics.

Agent programs

The agent programs all have the same skeleton: they take the current percept as input from the sensors and return an action to the actuatom6 Notice the difference between the agent program, which takes the current percept as input, and the agent function, which takes the entire percept history. The agent program takes just the current percept as input because nothing more is available from the environment; if the agent's actions depend on the entire percept sequence, the agent will have to remember the percepts.

Function TABLE-DRIVEN_AGENT(*percept*) **returns** an action
static: *percepts*, a sequence initially empty
table, a table of actions, indexed by percept sequence append
percept to the end of *percepts*
action □ LOOKUP(*percepts*, *table*)
return *action*

Figure 1.8 The TABLE-DRIVEN-AGENT program is invoked for each new percept and returns an action each time.

Drawbacks

- Table lookup of percept-action pairs defining all possible condition-action rules necessary to interact in an environment
- Problems
    1. Too big to generate and to store (Chess has about 10^120 states, for example)
    2. No knowledge of non-perceptual parts of the current state
    3. Not adaptive to changes in the environment; requires entire table to be updated if changes occur
    4. Looping: Can't make actions conditional

- Take a long time to build the table
- No autonomy
- Even with learning, need a long time to learn the table entries

1.3. Typical Intelligent Agents

- **Table-driven agents**

1. use a percept sequence/action table in memory to find the next action. They are implemented by a (large) **lookup table**.

- **Simple reflex agents**

1. are based on **condition-action rules**, implemented with an appropriate production system. They are stateless devices which do not have memory of past world states.

- **Agents with memory**

1. have **internal state**, which is used to keep track of past states of the world.

- **Agents with goals**

1. are agents that, in addition to state information, have **goal information** that describes desirable situations. Agents of this kind take future events into consideration.

- **Utility-based agents**

1. base their decisions on **classic axiomatic utility theory** in order to act rationally

Simple Reflex Agent

The simplest kind of agent is the **simple reflex agent.** These agents select actions on the basis of the *current* percept, ignoring the rest of the percept history. For example, the vacuum agent whose agent function is tabulated in Figure 1.10 is a simple reflex agent, because its decision is based only on the current location and on whether that contains dirt.

- Select action on the basis of *only the current* percept.
- E.g. the vacuum-agent
- Large reduction in possible percept/action situations (next page).
- Implemented through *condition-action rules.* If dirty then suck.

A Simple Reflex Agent: Schema

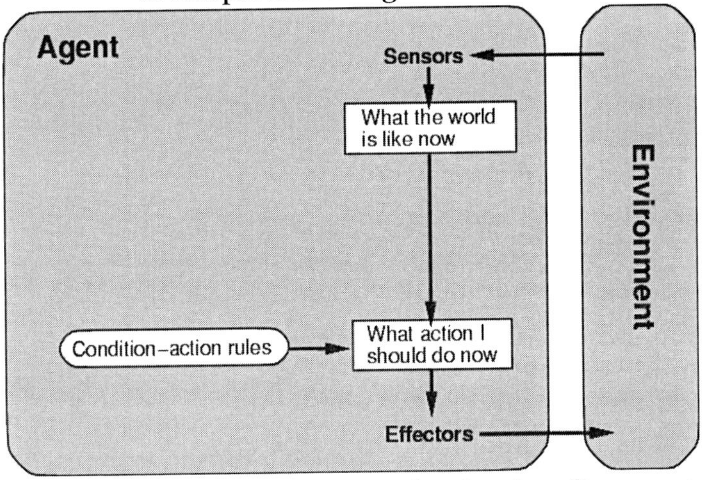

Figure 1.9 Schematic diagram of a simple reflex agent

function SIMPLE-REFLEX-AGENT(*percept*) **returns** an action **static**: *rules*, a set of condition-action rules *state* □ INTERPRET-INPUT(*percept*) *rule* □ RULE-MATCH(*state*, *rule*) *action* □ RULE-ACTION[*rule*]return *action*
Figure 1.10 A simple reflex agent. It acts according to a rule whose condition matches the current state, as defined by the percept
function REFLEX-VACUUM-AGENT ([*location*, *status*]) return

```
an action
if status == Dirty then return Suck
else if location == A then return Right
else if location == B then return Left
```

Figure 1.11 The agent program for a simple reflex agent in the two-state vacuum environment. This program implements the agent function tabulated in the figure 1.4.

Characteristics

- Only works if the environment is fully observable.
- Lacking history, easily get stuck in infinite loops
- One solution is to randomize actions

Model-based reflex agents

The most effective way to handle partial observability is for the agent to keep track of the part of the world it can't see now. That is, the agent should maintain some sort of internal state that depends on the percept history and thereby reflects at least some of the unobserved aspects of the current state.

Updating this internal state information as time goes by requires two kinds of knowledge to be encoded in the agent program. First, we need some information about how the world evolves independently of the agent-for example, that an overtaking car generally will be closer behind than it was a moment ago. Second, we need some information about how the agent's own actions affect the world-for example, that when the agent turns the steering wheel clockwise, the car turns to the right or that after driving for five minutes northbound on the freeway one is usually about five miles north of where one was five minutes ago. This knowledge about "how the world working - whether implemented in simple Boolean circuits or in complete scientific theories-is called a model of the world. An agent that uses such a MODEL-BASED model is called a model-based agent.

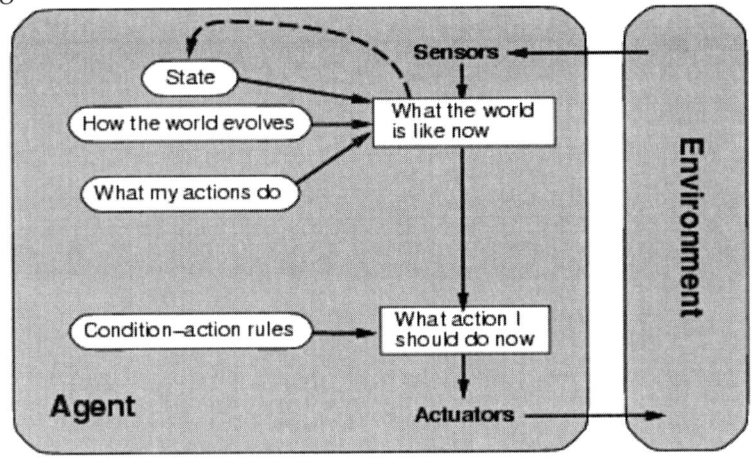

Figure 1.12 A model based reflex agent

function REFLEX-AGENT-WITH-STATE(percept) returns an action static: rules, a set of condition-action rules
state, a description of the current world state
action, the most recent action.
state □ UPDATE-STATE(state, action, percept)
rule □ RULE-MATCH(state, rule) action □ RULE-ACTION[rule] return action

Figure 1.13 Model based reflex agent. It keeps track of the current state of the world using an internal model. It then chooses an action in the same way as the reflex agent

Goal-based agents

Knowing about the current state of the environment is not always enough to decide what to do. For example, at a road junction, the taxi can turn left, turn right, or go straight on. The correct decision depends on where the taxi is trying to get to.

In other words, as well as a current state description, the agent needs some sort of goal information that describes situations that are desirable-for example, being at the passenger's destination. The agent program can combine this with information about the results of possible actions (the same information as was used to update internal state in the reflex agent) in order to choose actions that achieve the goal. Figure 1.13 shows the goal-based agent's structure.

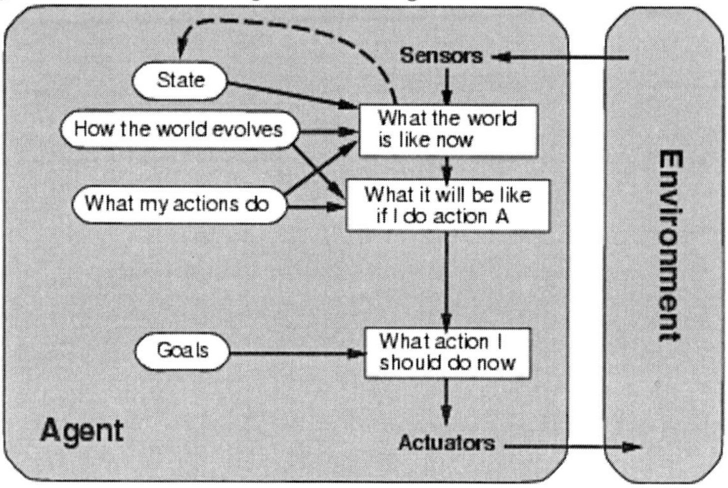

Figure 1.14 A goal based agent

Utility-based agents

Goals alone are not really enough to generate high-quality behavior in most environments. For example, there are many action sequences that will get the taxi to its destination (thereby achieving the goal) but some are quicker, safer, more reliable, or cheaper than others. Goals just provide a crude binary distinction between "happy" and "unhappy" states, whereas a more general performance measure should allow a comparison of different world states according to exactly how happy they would make the agent if they could be achieved.

Because "happy" does not sound very scientific, the customary terminology is to say that if one world state is preferred to another, then it has higher utility for the agent.

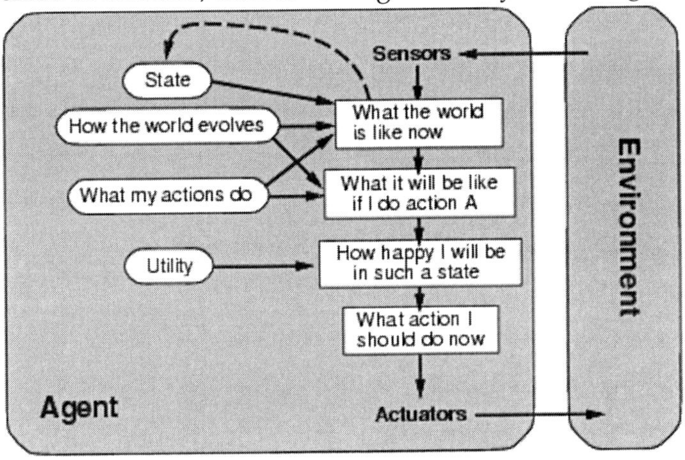

Figure 1.15 A model-based, utility-based agent.

It uses a model of the world, along with a utility function that measures its preferences among states of the world. Then it chooses the action that leads to the best expected utility, where expected utility is computed by averaging over all possible outcome states, weighted by the probability of the outcome.

1. Certain goals can be reached in different ways.
   - Some are better, have a higher utility.
2. Utility function maps a (sequence of) state(s) onto a real number.
3. Improves on goals:
   - Selecting between conflicting goals
   - Select appropriately between several goals based on likelihood of success.
4. All agents can improve their performance through learning.

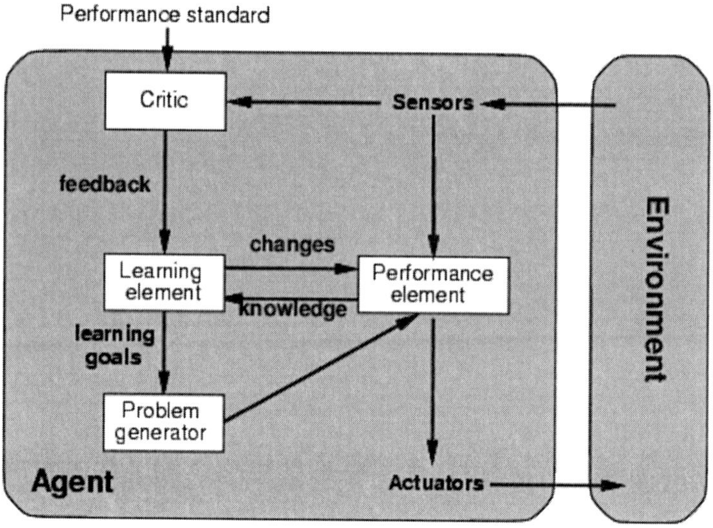

Figure 1.16 A general model of learning agents.

A learning agent can be divided into four conceptual components, as shown in Figure 1.15 The most important distinction is between the learning element, which is responsible for making improvements, and the performance element, which is responsible for selecting external actions. The performance element is what we have previously considered to be the entire agent: it takes in percepts and decides on actions. The learning element uses feedback from the critic on how the agent is doing and determines how the performance element should be modified to do better in the future.

The last component of the learning agent is the problem generator. It is responsible for suggesting actions that will lead to new and informative experiences. But if the agent is willing to explore a little, it might discover much better actions for the long run. The problem generator's job is to suggest these exploratory actions. This is what scientists do when they carry out experiments.

Summary: Intelligent Agents

- An agent perceives and acts in an environment, has an architecture, and is implemented by an agent program.
- Task environment – PEAS (Performance, Environment,
- Actuators, Sensors)
- The most challenging environments are inaccessible, nondeterministic, dynamic, and continuous.
- An ideal agent always chooses the action which maximizes its expected performance, given its percept sequence so far.
- An agent program maps from percept to action and updates internal state.
- Reflex agents respond immediately to percepts.
  1. simple reflex agents
  2. model-based reflex agents
- Goal-based agents act in order to achieve their goal(s).
- Utility-based agents maximize their own utility function.
- All agents can improve their performance through learning.

1.4. Problem Solving Approach to Typical AI problems

Problem Solving by Search

An important aspect of intelligence is *goal-based* problem solving.

The solution of many problems can be described by finding a sequence of actions that lead to a desirable goal. Each action changes the *state* and the aim is to find the sequence of actions and states that lead from the initial (start) state to a final (goal) state.

A well-defined problem can be described by:

- Initial state

- Operator or successor function - for any state x returns s(x), the set of states reachable from x with one action
- State space - all states reachable from initial by any sequence of actions
- Path - sequence through state space
- Path cost - function that assigns a cost to a path. Cost of a path is the sum of costs of individual actions along the path
- Goal test - test to determine if at goal state

What is Search?

Search is the systematic examination of states to find path from the start/root state to the goal state.

The set of possible states, together with *operators* defining their connectivity constitute the *search space.*

The output of a search algorithm is a solution, that is, a path from the initial state to a state that satisfies the goal test.

Problem-solving agents

A Problem solving agent is a goal-based agent. It decide what to do by finding sequence of actions that lead to desirable states. The agent can adopt a goal and aim at satisfying it.

To illustrate the agent's behavior, let us take an example where our agent is in the city of Arad, which is in Romania. The agent has to adopt a goal of getting to Bucharest.

Goal formulation, based on the current situation and the agent's performance measure, is the first step in problem solving. The agent's task is to find out which sequence of actions will get to a goal state. Problem formulation is the process of deciding what actions and states to consider given a goal.

Example: Route finding problem
Referring to figure 1.19
On holiday in Romania: currently in Arad. Flight leaves tomorrow from Bucharest **Formulate goal**: be in Bucharest

Formulate problem states: various cities
actions: drive between cities

Find solution
sequence of cities, e.g., Arad, Sibiu, Fagaras, Bucharest
Problem formulation

A **problem** is defined by four items:
initial state e.g., –at Arad"
successor function S(x) = set of action-state pairs e.g., S(Arad) = {[Arad -> Zerind;Zerind],....}
goal test, can be
explicit, e.g., x = at Bucharest" implicit, e.g., NoDirt(x)
path cost (additive)
e.g., sum of distances, number of actions executed, etc. c(x; a; y) is the step cost, assumed to be >= 0
A **solution** is a sequence of actions leading from the initial state to a goal state.

Figure 1.17 Goal formulation and problem formulation

Search

An agent with several immediate options of unknown value can decide what to do by examining different possible sequences of actions that leads to the states of known value, and then choosing the best sequence. The process of looking for sequences actions from the current state to reach the goal state is called search.

The search algorithm takes a problem as input and returns a solution in the form of action sequence. Once a solution is found, the execution phase consists of carrying out the recommended action.

Figure 1.18 shows a simple –formulate, search, executeǁ design for the agent. Once solution has been executed, the agent will formulate a new goal.

function SIMPLE-PROBLEM-SOLVING-AGENT(*percept*)
returns an action
inputs : *percept*, a percept
static: *seq*, an action sequence, initially empty *state*, some description of the current world state *goal*, a goal, initially null *problem*, a problem formulation
state UPDATE-STATE(*state, percept*)
if seq is empty **then do**
goal FORMULATE-GOAL(*state*)
problem FORMULATE-PROBLEM(*state, goal*)
seq SEARCH(*problem*)
action FIRST(*seq*);
seq REST(seq)
return *action*

Figure 1.18 A Simple problem solving agent. It first formulates a goal and a problem, searches for a sequence of actions that would solve a problem, and executes the actions one at a time.

The agent design assumes the Environment is

- **Stati**: The entire process carried out without paying attention to changes that might be occurring in the environment.

- **Observable:** The initial state is known and the agent's sensor detects all aspects that are relevant to the choice of action
- **Discrete:** With respect to the state of the environment and percepts and actions so that alternate courses of action can be taken
- **Deterministic:** The next state of the environment is completely determined by the current state and the actions executed by the agent. Solutions to the problem are single sequence of actions

An agent carries out its plan with eye closed. This is called an open loop system because ignoring the percepts breaks the loop between the agent and the environment.

1.3.1.1 Well-defined problems and solutions

A problem can be formally defined by four components:

- The initial state that the agent starts in. The initial state for our agent of example problem is described by In(Arad)
- A Successor Function returns the possible actions available to the agent. Given a state x, SUCCESSOR- FN(x) returns a set of {action, successor} ordered pairs where each action is one of the legal actions in state x, and each successor is a state that can be reached from x by applying the action.

For example, from the state In (Arad), the successor function for the Romania problem would return
{[Go(Sibiu),In(Sibiu)],[Go(Timisoara),In(Timisoara)],
[Go(Zerind),In(Zerind)] }

- State Space: The set of all states reachable from the initial state. The state space forms a graph in which the nodes are states and the arcs between nodes are actions.
- A path in the state space is a sequence of states connected by a sequence of actions.

- The goal test determines whether the given state is a goal state.
- A path cost function assigns numeric cost to each
- action. For the Romania problem the cost of path might be its length in kilometers.
- The step cost of taking action a to go from state x to state y is denoted by c(x, a, y). The step costs for Romania are shown in figure 1.18. It is assumed that the step costs are non-negative.
- A solution to the problem is a path from the initial state to a goal state.
- An optimal solution has the lowest path cost among all solutions.

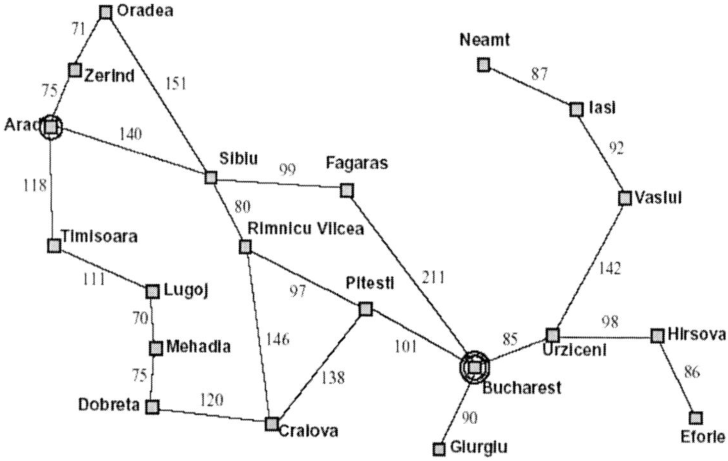

Figure 1.19 A simplified Road Map of part of Romania

1.3.2 Example Problems

The problem solving approach has been applied to a vast array of task environments. Some best known problems are ummarized below. They are distinguished as toy or real- world problems a toy problem is intended to illustrate various problem solving methods. It can be easily used by different researchers to compare the performance of algorithms.

A real world problem is one whose solutions people actually care about.

1.3.2.1 Toy Problems

Vacuum World Example

- States: The agent is in one of two locations. Each of which might or might not contain dirt. Thus there are 2 x 22 = 8 possible world states.
- Initial state: Any state can be designated as initial state.
- Successor function: This generates the legal states that results from trying the three actions (left, right, suck). The complete state space is shown in figure 2.3
- Goal Test: This tests whether all the squares are clean.
- Path test: Each step costs one, so that the the path cost is the number of steps in the path.

Vacuum World State Space

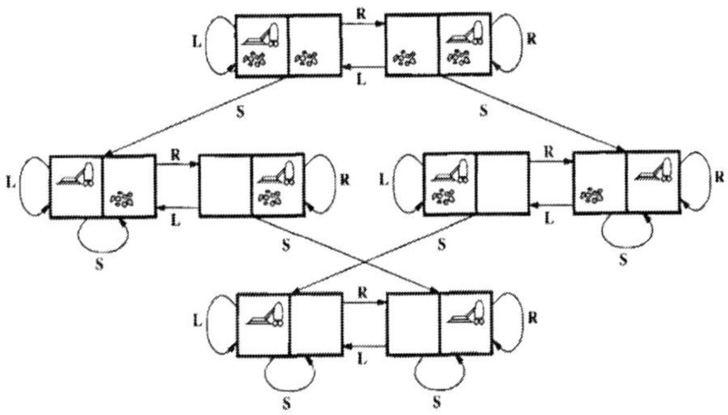

Figure 1.20 The state space for the vacuum world. Arcs denote actions: L = Left, R = Right, S = Suck

The 8-puzzle

An 8-puzzle consists of a 3x3 board with eight numbered tiles and a blank space. A tile adjacent to the blank space can slide into the space. The object is to reach the goal state, as shown in figure 2.4

Example: The 8-puzzle

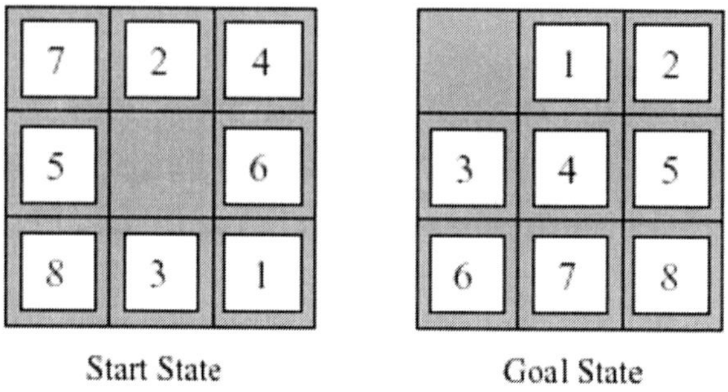

Figure 1.21 A typical instance of 8-puzzle.

The problem formulation is as follows

- **States**: A state description specifies the location of each of the eight tiles and the blank in one of the nine squares.
- **Initial state**: Any state can be designated as the initial state. It can be noted that any given goal can be reached from exactly half of the possible initial states.
- **Successor function**: This generates the legal states that result from trying the four actions (blank moves Left, Right, Up or down).
- **Goal Test** : This checks whether the state matches the goal configuration shown in figure 2.4. (Other goal configurations are possible)
- **Path cost**: Each step costs 1, so the path cost is the number of steps in the path.

- **The 8-puzzle** belongs to the family **of sliding-block puzzles,** which are often used as test problems for new search algorithms in AI. This general class is known as

NP-complete.

- The **8-puzzle** has 9!/2 = 181,440 reachable states and is
- easily solved.
- The **15 puzzle** (4 x 4 board) has around 1.3 trillion states, an the random instances can be solved optimally in few milli seconds by the best search algorithms.
- The **24-puzzle** (on a 5 x 5 board) has around 10^{25} states, and random instances are still quite difficult to solve optimally with current machines and algorithms.

8-queens problem

The goal of 8-queens problem is to place 8 queens on the chessboard such that no queen attacks any other. (A queen attacks any piece in the same row, column or diagonal).

Figure 2.5 shows an attempted solution that fails: the queen in the right most column is attacked by the queen at the top left.

Figure 1.22 8-queens problem

An **Incremental formulation** involves operators that augment the state description, starting with an empty state. For 8-queens problem, this means each action adds a queen to the state.

A **complete-state formulation** starts with all 8 queens on the board and moves them around.

In either case the path cost is of no interest because only the final state counts.

The first incremental formulation one might try is the following:

- States: Any arrangement of 0 to 8 queens on board is a state.
- Initial state: No queen on the board.
- Successor function: Add a queen to any empty square.
- Goal Test: 8 queens are on the board, none attacked.

In this formulation, we have 64.63…57 = 3 x 1014 possible sequences to investigate.

A better formulation would prohibit placing a queen in any square that is already attacked. :

- States: Arrangements of n queens (0 <= n < = 8), one per column in the left most columns ,with no queen attacking another are states.
- Successor function: Add a queen to any square in the left most empty column such that it is not attacked by any other queen.

This formulation reduces the 8-queen state space from 3 x 1014 to just 2057,and solutions are easy to find.

For the 100 queens the initial formulation has roughly 10400 states whereas the improved formulation has about 1052 states. This is a huge reduction, but the improved state space is still too big for the **algorithms to handle.**

1.3.2.2 Real-World Problems

Route - Finding Problem

Route-finding problem is defined in terms of specified locations and transitions along links between them. Route-finding algorithms are used in a variety of applications, such as routing in computer networks, military operations planning, and air line travel planning systems.

Airline Travel Problem

The **airline travel problem** is specifies as follows:

- **States:** Each is represented by a location(e.g.,an airport) and the current time.
- **Initial state:** This is specified by the problem.
- **Successor function:** This returns the states resulting from taking any scheduled flight (further specified by seat class and location), leaving later than the current time plus the within-airport transit time,from the current airport to another.
- **Goal Test:** Are we at the destination by some pre specified time?
- **Path cost:** This depends upon the monetary cost, waiting time, flight time, customs and immigration procedures, seat quality, time of dat, type of air plane, frequent-flyer mileage awards, and so on.

Touring Problems

Touring problems are closely related to route-finding problems, but with an important difference.

Consider for example, the problem‖, Visit every city at least once‖ as shown in Romania map.

As with route-finding the actions correspond to trips between adjacent cities. The state space, however, is quite different.

The initial state would be –In Bucharest; visited {Bucharest}‖.

A typical intermediate state would be –In Vaslui; visited

{Bucharest, Urziceni, Vaslui }‖.

The goal test would check whether the agent is in Bucharest and all 20 cities have been visited.

The Travelling Salesperson Problem (TSP)

Is a touring problem in which each city must be visited exactly once. The aim is to find the shortest tour. The problem is known to be **NP-hard**. Enormous efforts have been expended to improve the capabilities of TSP algorithms. These algorithms are also used in tasks such as planning movements of **automatic circuit-board drills** and of **stocking machines** on shop floors.

VLSI layout

A **VLSI layout** problem requires positioning millions of components and connections on a chip to minimize area, minimize circuit delays, minimize stray capacitances, and maximize manufacturing yield. The layout problem is split into two parts: **cell layout** and **channel routing**.

ROBOT navigation

ROBOT navigation is a generalization of the route-finding problem. Rather than a discrete set of routes, a robot can move in a continuous space with an infinite set of possible actions and states. For a circular Robot moving on a flat surface, the space is essentially two-dimensional.

When the robot has arms and legs or wheels that also must be controlled, the search space becomes multi-dimensional. Advanced techniques are required to make the search space finite.

Automatic Assembly Sequencing

The example includes assembly of intricate objects such as electric motors. The aim in assembly problems is to find the order in which to assemble the parts of some objects. If the wrong order is chosen, there will be no way to add some part later without undoing some work already done.

Another important assembly problem is protein design, in which the goal is to find a sequence of Amino acids that will be fold into a three dimensional protein with the right properties to cure some disease.

Internet Searching

In recent years there has been increased demand for software robots that perform Internet searching, looking for answers to questions, for related information or for shopping deals. The searching techniques consider internet as a graph of nodes (pages) connected by links.

1.3.2 Searching For Solutions

Search Tree

Having formulated some **problems**, we now need to **solve** them. This is done by a **search** through the **state space**. A **search tree** is generated by the **initial state** and the **successor function** that together define the **state space**. In general, we may have a *search graph* rather than a *search tree*, when the same state can be reached from multiple paths.

Figure 1.23 shows some of the expansions in the search tree for finding a route from Arad to Bucharest.

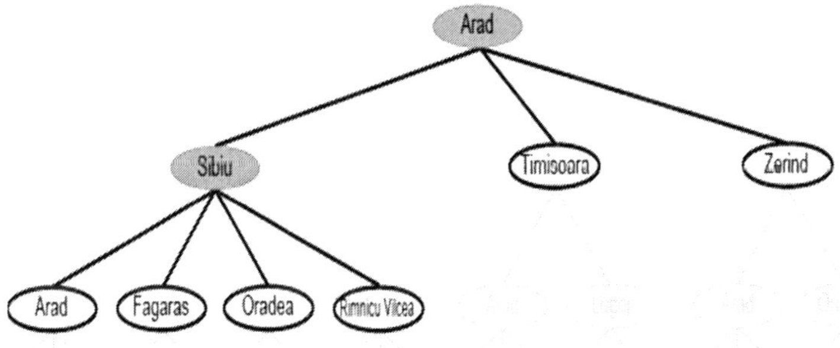

Figure 1.23

Partial search trees for finding a route from Arad to Bucharest. Nodes that have been expanded are shaded.; nodes that have been generated but not yet expanded are outlined in bold; nodes that have not yet been generated are shown in faint dashed line The root of the search tree is a **search node** corresponding to the initial **state**, In (Arad). The first step is to test whether this is a **goal state**. The current state is expanded by applying the successor function to the current state, thereby generating a new set of states. In this case,we get three new states: In(Sibiu),In(Timisoara),and In(Zerind). Now we must choose which of these three possibilities to consider further. This is the essense of search- following up one option now and putting the others aside for latter,in case the first choice does not lead to a solution.

Search strategy. The general tree-search algorithm is described informally in Figure 1.24.

Tree Search

The choice of which state to expand is determined by the **search strategy**. There are infinite number paths in this state space, so the search tree has an infinite number of **nodes.**

A **node** is a data structure with five components:

- STATE : a state in the state space to which the node corresponds;
- PARENT-NODE : the node in the search tree that generated this node;
- ACTION : the action that was applied to the parent to generate the node;
- PATH-COST :the cost,denoted by g(n),of the path from initial state to the node,as indicated by the parent pointers; and
- DEPTH : the number of steps along the path from the initial state.

It is important to remember the distinction between nodes and states. A node is a book keeping data structure used to represent the search tree. A state corresponds to configuration of the world.

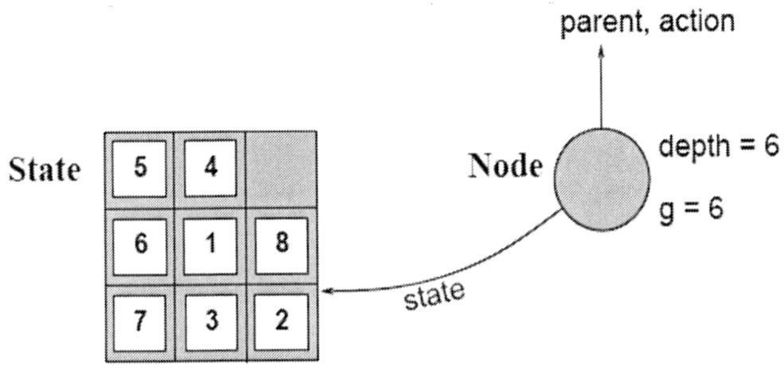

Figure 1.25 Nodes are data structures from which the search tree is constructed. Each has a parent, a state, Arrows point from child to parent

Fringe

Fringe is a collection of nodes that have been generated but not yet been expanded. Each element of the fringe is a leaf node, that is, a node with no successors in the tree. The fringe of each tree consists of those nodes with bold outlines. The collection of these nodes is implemented as a queue.

The general tree search algorithm is shown in Figure 2.9

```
function TREE-SEARCH(problem, fringe) returns a solution, or failure

  fringe ← INSERT(MAKE-NODE(INITIAL-STATE[problem]), fringe)
  loop do
      if EMPTY?(fringe) then return failure
      node ← REMOVE-FIRST(fringe)
      if GOAL-TEST[problem] applied to STATE[node] succeeds
          then return SOLUTION(node)
      fringe ← INSERT-ALL(EXPAND(node, problem), fringe)
```

```
function EXPAND(node, problem) returns a set of nodes
  successors ← the empty set
  for each ⟨action, result⟩ in SUCCESSOR-FN[problem](STATE[node]) do
    s ← a new NODE
    STATE[s] ← result
    PARENT-NODE[s] ← node
    ACTION[s] ← action
    PATH-COST[s] ← PATH-COST[node] + STEP-COST(node, action, s)
    DEPTH[s] ← DEPTH[node] + 1
    add s to successors
  return successors
```

Figure 1.26 The general Tree search algorithm

The operations specified in Figure 1.26 on a queue are as follows:

- **MAKE-QUEUE (element,...)** creates a queue with the given element(s).
- **EMPTY? (queue)** returns true only if there are no more
- elements in the queue.
- **FIRST(queue)** returns FIRST(queue) and removes it from the queue.
- **INSERT (element, queue)** inserts an element into the queue and returns the resulting queue.
- **INSERT-ALL (elements, queue)** inserts a set of elements into the queue and returns the resulting queue.

Measuring Problem-Solving Performance

The output of problem-solving algorithm is either failure or a solution. (Some algorithms might struck in an infinite loop and never return an output.

The algorithm's performance can be measured in four ways:

- **Completeness**: Is the algorithm guaranteed to find a solution when there is one?
- **Optimality** : Does the strategy find the optimal solution

- **Time complexity:** How long does it take to find a solution?
- **Space complexity:** How much memory is needed to perform the search?

1.3.3 Uninformed Search Stratges

Uninformed Search Strategies have no additional information about states beyond that provided in the problem definition. Strategies that know whether one non goal state is more promising‖ than another are called informed search or heuristic search strategies.

There are five uninformed search strategies as given below.

- Breadth-first search
- Uniform-cost search
- Depth-first search
- Depth-limited search
- Iterative deepening search

1.3.4.1 Breadth-first search

Breadth-first search is a simple strategy in which the root node is expanded first, then all successors of the root node are expanded next, then their successors, and so on. In general, all the nodes are expanded at a given depth in the search tree before any nodes at the next level are expanded.

Breath-first-search is implemented by calling TREE-SEARCH with an empty fringe that is a first-in-first-out (FIFO) queue, assuring that the nodes that are visited first will be expanded first.

In other words, calling TREE-SEARCH (problem, FIFO-UEUE()) results in breadth-first-search.

The FIFO queue puts all newly generated successors at the end of the queue, which means that Shallow nodes are expanded before deeper nodes.

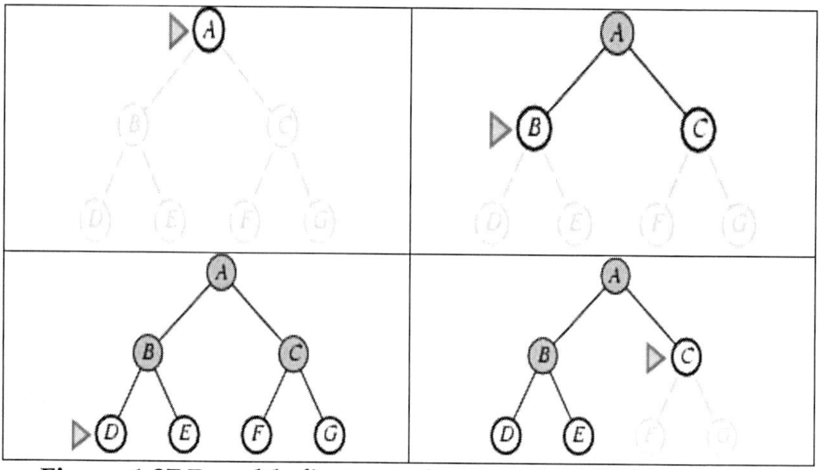

Figure 1.27 Breadth-first searches on a simple binary tree. At each stage, the node to be expanded next is indicated by a marker.

Properties of breadth-first-search

Complete?? Yes (if b is finite)

Time?? $1 + b + b^2 + b^3 + \ldots + b^d + b(b^d - 1) = O(b^{d+1})$, i.e., exp. in d

Space?? $O(b^{d+1})$ (keeps every node in memory)

Optimal?? No, unless step costs are constant

Space is the big problem; can easily generate nodes at 100MB/sec
so 24hrs = 8640GB.

Figure 1.28 Breadth-first-search properties

Time and Memory Requirements for BFS – $O(b^{d+1})$

Example:
- b = 10
- 10000 nodes/second
- each node requires 1000 bytes of storage

Depth	Nodes	Time	Memory
2	1100	.11 sec	1 meg
4	111,100	11 sec	106 meg
6	10^7	19 min	10 gig
8	10^9	31 hrs	1 tera
10	10^{11}	129 days	101 tera
12	10^{13}	35 yrs	10 peta
14	10^{15}	3523 yrs	1 exa

Figure 1.29

Time and memory requirements for breadth-first-search. The numbers shown assume branch factor of b = 10 ; 10,000 nodes/second; 1000 bytes/node

Time complexity for BFS

Assume every state has b successors. The root of the search tree generates b nodes at the first level,each of which generates b more nodes,for a total of b2 at the second level. Each of these generates b more nodes,yielding b3 nodes at the third level,and so on. Now suppose,that the solution is at depth d. In the worst case,we would expand all but the last node at level d,generating bd+1 - b nodes at level d+1.

Then the total number of nodes generated is b + b2 + b3 + ...+ bd + (bd+1 + b) – O(bd+1).

Every node that is generated must remain in memory, because it is either part of the fringe or is an ancestor of a fringe node. The space complexity is, therefore ,the same as the time complexity

1.3.4.2 Uniform-Cost Search

Instead of expanding the shallowest node, **uniform-cost search** expands the node n with the lowest path cost. Uniform- cost search does not care about the number of steps a path has, but only about their total cost.

Expand least-cost unexpanded node

Implementation:

fringe = queue ordered by path cost, lowest first

Equivalent to breadth-first if step costs all equal

Complete?? Yes, if step cost $\geq \epsilon$

Time?? # of nodes with $g \leq$ cost of optimal solution, $O(b^{\lceil C^*/\epsilon \rceil})$
where C^* is the cost of the optimal solution

Space?? # of nodes with $g \leq$ cost of optimal solution, $O(b^{\lceil C^*/\epsilon \rceil})$

Optimal?? Yes—nodes expanded in increasing order of $g(n)$

Figure 1.30 Properties of Uniform-cost-search

2.5.1.3 Depth-First-Search

Depth-first-search always expands the deepest node in the current fringe of the search tree. The progress of the search is illustrated in figure 1.31. The search proceeds immediately to the deepest level of the search tree, where the nodes have no successors. As those nodes are expanded, they are dropped from the fringe, so then the search –backs up‖ to the next shallowest node that still has unexplored successors.

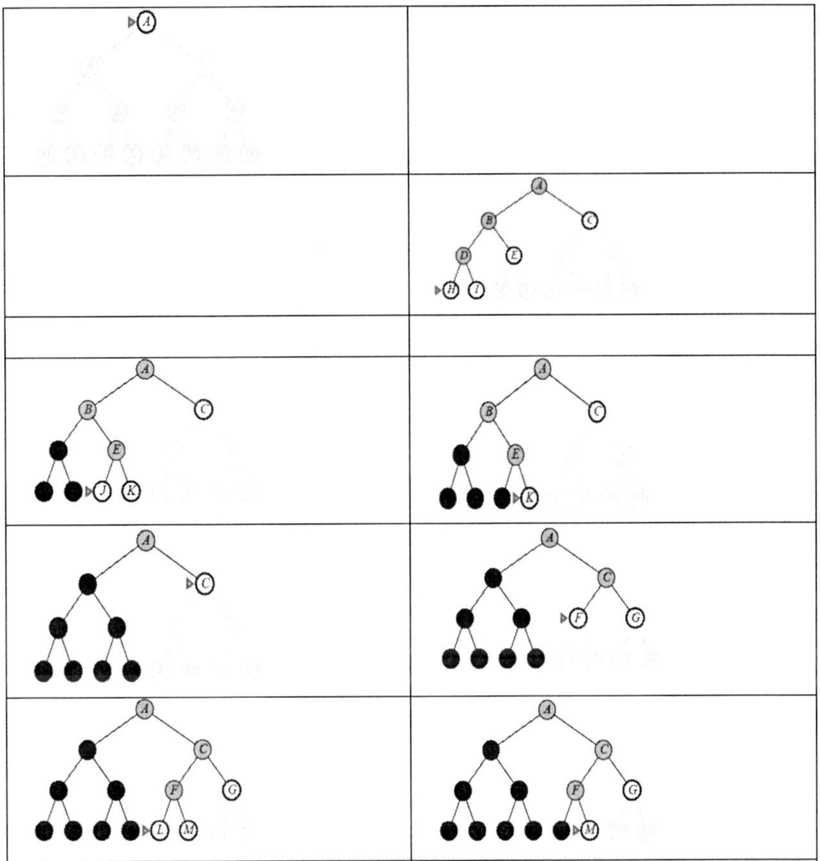

Figure 1.31 Depth-first-search on a binary tree. Nodes that have been expanded and have no descendants in the fringe can be removed from the memory; these are shown in black. Nodes at depth 3 are assumed to have no successors and M is the only goal node.

This strategy can be implemented by TREE-SEARCH with a last-in-first-out (LIFO) queue, also known as a stack.

Depth-first-search has very modest memory requirements. It needs to store only a single path from the root to a leaf node, along with the remaining unexpanded sibling nodes for each node on the path. Once the node has been expanded, it can be removed from the memory, as soon as its descendants have been fully explored (Refer Figure 2.12).

For a state space with a branching factor b and maximum depth m, depth-first-search requires storage of only bm + 1 nodes.

Using the same assumptions as Figure 2.11, and assuming that nodes at the same depth as the goal node have no successors, we find the depth-first-search would require 118 kilobytes instead of 10 petabytes, a factor of 10 billion times less space.

Drawback of Depth-first-search

The drawback of depth-first-search is that it can make a wrong choice and get stuck going down very long(or even infinite) path when a different choice would lead to solution near the root of the search tree. For example, depth-first-search will explore the entire left subtree even if node C is a goal node.

Backtracking Search

A variant of depth-first search called backtracking search uses less memory and only one successor is generated at a time rather than all successors.; Only O(m) memory is needed rather than O(bm)

1.3.4.4 Depth-Limited-Search

The problem of unbounded trees can be alleviated by supplying depth-first-search with a pre-determined depth limit l. That is nodes at depth l are treated as if they have no successors. This approach is called **depth-limited-search**. The depth limits solves the infinite path problem.

Depth limited search will be no optimal if we choose l > d.

Its time complexity is $O(b^l)$ and its space complexity is O(bl). Depth-first-search can be viewed as a special case of depth- limited search with l = oo

Sometimes, depth limits can be based on knowledge of the problem. For, example, on the map of Romania there are 20 cities. Therefore, we know that if there is a solution. it must be of length 19 at the longest, So l = 10 is a possible choice. However, it can be shown that any city can be reached from any other city in at most 9 steps. This number known as the **diameter** of the state space, gives us a better depth limit.

Depth-limited-search can be implemented as a simple modification to the general tree-search algorithm or to the recursive depth-first-search algorithm. The pseudo code for recursive depth-limited-search is shown in Figure 1.32.

It can be noted that the above algorithm can terminate with two kinds of failure: the standard *failure* value indicates no solution; the *cutoff* value indicates no solution within the depth limit.

Depth-limited search = depth-first search with depth limit l, returns cut off if any path is cut off by depth limit

function Depth-Limited-Search(problem, limit) **returns** a solution/
fail/cutoff
return Recursive-DLS(Make-Node(Initial-State[problem]), problem, limit)
function Recursive-DLS (node, problem, limit) returns solution/
fail/ cutoff
cutoff - occurred? false
if Goal-Test (problem, State[node]) then **return** Solution(node)

else if Depth[node] = limit **then return** cutoff
else for each successor **in** Expand(node, problem) **do**
result Recursive-DLS(successor, problem, limit)
if result = cutoff then cutoff_occurred? true
else if result not = failure **then return** result
if cutoff_occurred? then return cutoff **else return** failure

Figure 1.32 Recursive implementation of Depth-limited-search

1.3.4.5 Iterative Deepening Depth-First Search

Iterative deepening search (or iterative-deepening-depth-first-search) is a general strategy often used in combination with depth-first-search, that finds the better depth limit. It does this by gradually increasing the limit – first 0, then 1, then 2, and so on – until a goal is found. This will occur when the depth limit reaches d,the depth of the shallowest goal node. The algorithm is shown in Figure 2.14.

Iterative deepening combines the benefits of depth-first and breadth-first-search Like depth-first-search, its memory requirements are modest; O(bd) to be precise.

Like Breadth-first-search, it is complete when the branching factor is finite and optimal when the path cost is a non- decreasing function of the depth of the node.

Figure 2.15 shows the four iterations of ITERATIVE-DEEPENING_SEARCH on a binary search tree, where the solution is found on the fourth iteration.

```
function ITERATIVE-DEEPENING-SEARCH(problem) returns a solution
   inputs: problem, a problem
   for depth ← 0 to ∞ do
      result ← DEPTH-LIMITED-SEARCH(problem, depth)
      if result ≠ cutoff then return result
   end
```

The **iterative deepening search algorithm,** which repeatedly applies depth-limited-search with increasing limits. It terminates when a solution is found or if the depth limited search returns failure, meaning that no solution exists.

Iterative deepening search

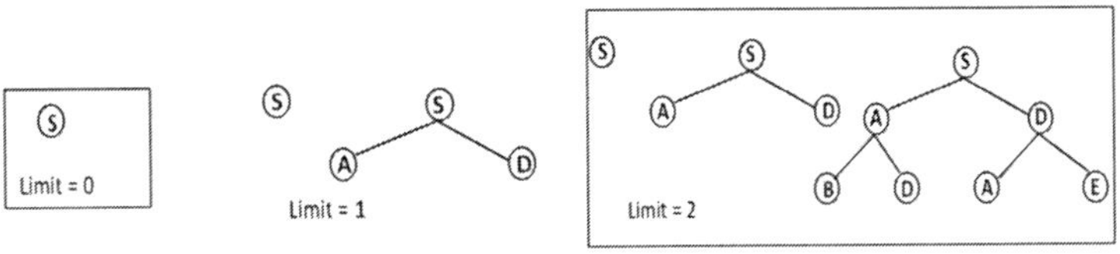

Figure 1.35

Iterative search is not as wasteful as it might seem

In general, iterative deepening is the prefered uninformed search method when there is a large search space and the depth of solution is not known.

1.3.4.6 Bidirectional Search

The idea behind bidirectional search is to run two simultaneous searches one forward from he initial state and the other backward from the goal, stopping when the two searches meet in the middle (Figure 1.37).

The motivation is that $b^{d/2} + b^{d/2}$ much less than, or in the figure, the area of the two small circles is less than the area of one big circle centered on the start and reaching to the goal.

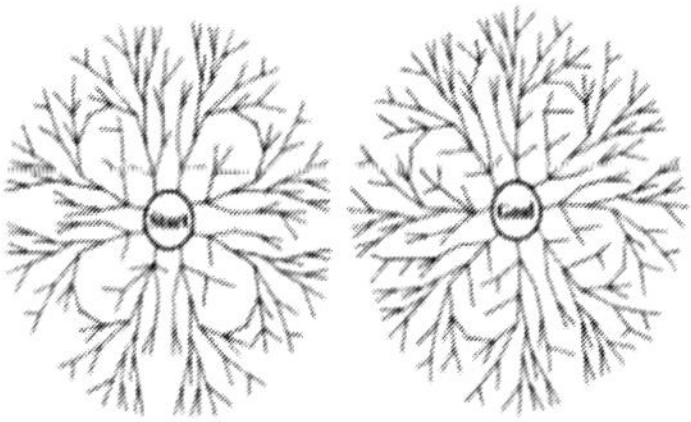

Figure 1.37

A schematic view of a bidirectional search that is about to succeed, when a Branch from the Start node meets a Branch from the goal node.

1.3.2 Avoiding Repeated States

In searching, time is wasted by expanding states that have already been encountered and expanded before. For some problems repeated states are unavoidable. The search trees for these problems are infinite. If we prune some of the repeated states, we can cut the search tree down to finite size. Considering search tree upto a fixed depth, eliminating repeated states yields an exponential reduction in search cost.

Repeated states, can cause a solvable problem to become unsolvable if the algorithm does not detect them.

Repeated states can be the source of great inefficiency: identical sub trees will be explored many times!

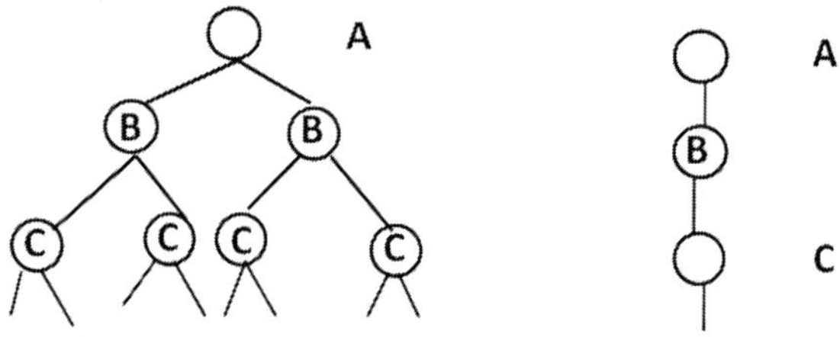

Figure 1.39 Avoiding Repatead States

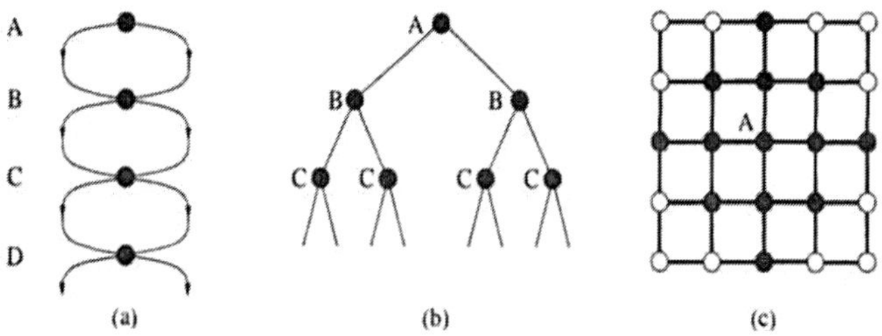

Figure 1.40

State space that generate an exponentially larger search tree. (a) A state space in which there are two possible actions leading from A to B. the state space contains d + 1 states, where d is the maximum depth. (b) search tree, which has 2^d branches corresponding to the 2d paths through rectangular grid space. States within 2 step initial state (A) are

Shown

- Do not return to the previous state.
- Do not create paths with cycles.
- Do not generate the same state twice.
- Store states in a hash table.
- Check for repeated states.
- Using more memory in order to check repeated state
- Algorithms that forget their history are doomed to repeat it.
- Maintain Close-List beside Open-List(fringe)

Strategies for avoiding repeated states

We can modify the general TREE-SEARCH algorithm to include the data structure called the closed list, which stores every expanded node. The fringe of unexpanded nodes is called the open list.

If the current node matches a node on the closed list, it is discarded instead of being expanded.

The new algorithm is called GRAPH-SEARCH and much more efficient than TREE-SEARCH. The worst case time and space requirements may be much smaller than O(bd).

1.3.4 Earching with Partial Information

Different types of incompleteness lead to three distinct problem types:

- Sensor less problems (conformant): If the agent has no sensors at all
- Contingency problem: if the environment if partially observable or if action are uncertain (adversarial)
- Exploration problems: When the states and actions of the environment are unknown.
- No sensor

- Initial State(1,2,3,4,5,6,7,8)
- After action [Right] the state (2,4,6,8)
- After action [Suck] the state (4, 8)
- After action [Left] the state (3,7)
- After action [Suck] the state (8)
- Answer : [Right, Suck, Left, Suck] coerce the world into state 7 without any sensor
  Belief State: Such state that agent belief to be there (SLIDE 7) Partial knowledge of states and actions:
- Sensor less or conformant problem
- Agent may have no idea where it is; solution (if any) is a sequence.
- contingency problem
- Percepts provide new information about current state; solution is a tree or policy; often interleave search and execution.
- If uncertainty is caused by actions of another agent:
- adversarial problem
- exploration problem
- When states and actions of the environment are unknown.

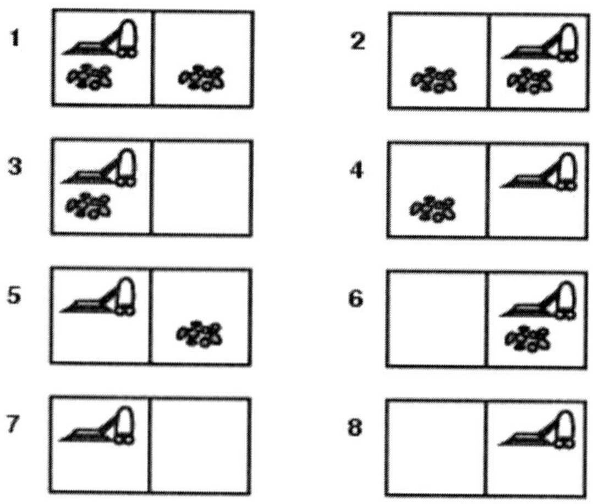

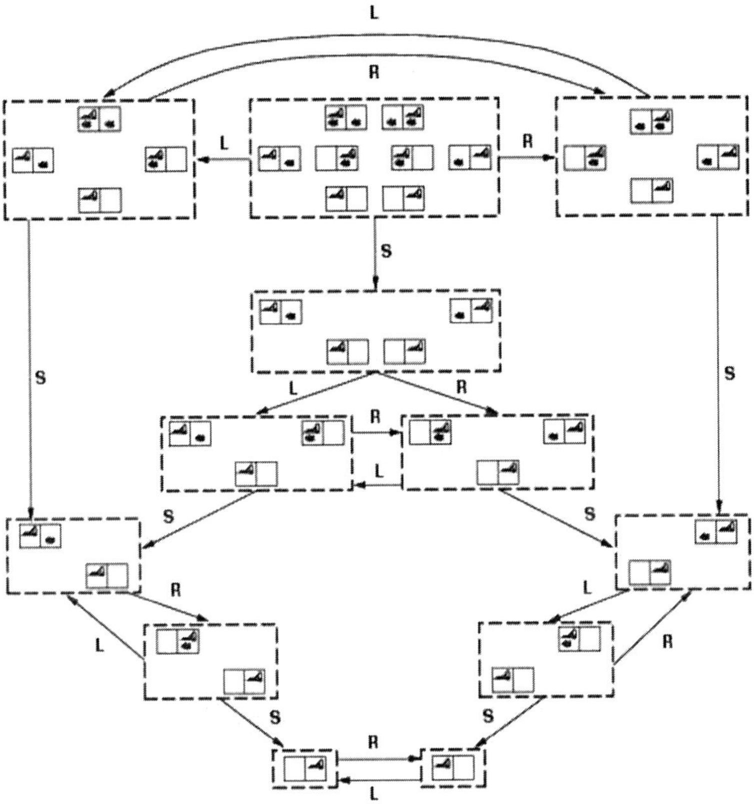

Contingency, start in {1,3}.

Murphy's law, Suck can dirty a clean carpet. Local sensing: dirt, location only.

- Percept = [L,Dirty] ={1,3}
- [Suck] = {5,7}
- [Right] ={6,8}
- [Suck] in {6}={8} (Success)
- BUT [Suck] in {8} = failure
- Solution??
- Belief-state: no fixed action sequence guarantees solution

- Relax requirement:

- [Suck, Right, if [R,dirty] then Suck]
- Select actions based on contingencies arising during execution.

Time and space complexity are always considered with respect to some measure of the problem difficulty. In theoretical computer science, the typical measure is the size of the state space.

In AI, where the graph is represented implicitly by the initial state and successor function, the complexity is expressed in terms of three quantities:

b, the branching factor or maximum number of successors of any node;

d, the depth of the shallowest goal node; and

m, the maximum length of any path in the state space. Search-cost - typically depends upon the time complexity but can also include the term for memory usage.

Total–cost – It combines the search-cost and the path cost of the solution found.

1 – QUESTION BANK

1. What is AI?

Systems that think like humans	Systems that think rationally
Systems that act like humans	Systems that act rationally

2. Define an agent.

An agent is anything that can be viewed as perceiving its environment through sensors and acting upon that environment through actuators.

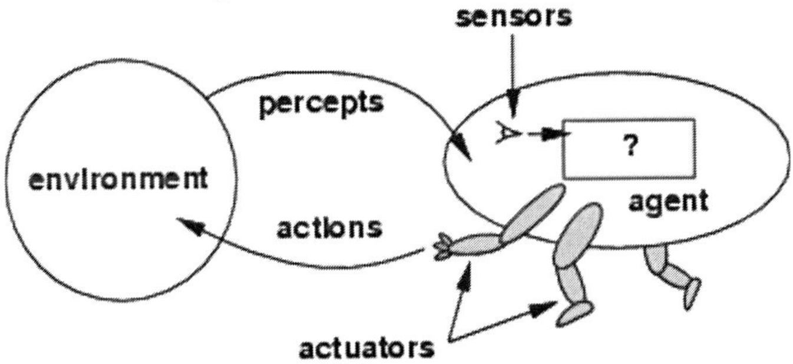

3. What is an agent function?

An agent's behavior is described by the agent function that maps any given percept sequence to an action.

4. Differentiate an agent function and an agent program.

Agent Function	Agent Program
An abstract mathematical description	A concrete implementation, running on the agent Architecture.

5. What is a task environment? How it is specified?

Task environments are essentially the "problems" to which rational agents are the "solutions."

A Task environment is specified using PEAS (Performance, Environment, Actuators, Sensors) description

6. List the properties of task environments.
   - Fully observable vs. partially observable.
   - Deterministic vs. stochastic.
   - Episodic vs. sequential
   - Static vs, dynamic.
   - Discrete vs. continuous.
   - single agent vs. multiagent.

7. What are the four different kinds of agent programs?
   - Simple reflex agents;
   - Model-based reflex agents;
   - Goal-based agents; and
   - Utility-based agents.

8. What are utility based agents?

Goals alone are not really enough to generate high-quality behavior in most environments.

For example, there are many action sequences that will get the taxi to its destination (thereby achieving the goal) but some are quicker, safer, more reliable, or cheaper than others.

A utility function maps a state (or a sequence of states) onto a real number, which describes the associated degree of happiness.

9. Define the problem solving agent.

A Problem solving agent is a goal-based agent. It decide what to do by finding sequence of actions that lead to desirable states. The agent can adopt a goal and aim at satisfying it.

Goal formulation is the first step in problem solving.

10. Define the terms goal formulation and problem formulation.

Goal formulation, based on the current situation and the agent's performance measure, is the first step in problem solving.

The agent's task is to find out which sequence of actions will get to a goal state.

Problem formulation is the process of deciding what actions and states to consider given a goal.

11. List the steps involved in simple problem solving agent.
    a. Goal formulation
    b. Problem formulation c. Search
    c. Search Algorithm
    d. Execution phase

12. Define search and search algorithm.

The process of looking for sequences actions from the current state to reach the goal state is called search.

The search algorithm takes a problem as input and returns a solution in the form of action sequence. Once a solution is found, the execution phase consists of carrying out the recommended action.

13. What are the components of well-defined problems?
    a. The initial state that the agent starts in. The initial state for our agent of example problem is described by In(Arad)
    b. A Successor Function returns the possible actions available to the agent. Given a state x,SUCCESSOR-FN(x) returns a set of {action, successor} ordered pairs where each action is one of the legal actions in state x,and each successor is a state that can be reached from x by applying the action.

For example, from the state In(Arad),the successor function for the Romania problem would return
{ [Go(Sibiu),In(Sibiu)],[Go(Timisoara),In(Timisoara)],[Go(Zerind),In(Zerind)] }

- The goal test determines whether the given state is a goal state.
- A path cost function assigns numeric cost to each action.
- For the Romania problem the cost of path might be its length in kilometers.

14. Differentiate toy problems and real world problems.

Toy problems	Real world problems
A toy problem is intended to illustrate various problem solving methods. It can be easily used by different researchers to compare the performance of algorithms	A real world problem is one whose solutions people actually care about.

15. Give examples of real world problems.
    - Touring problems
    - Travelling Salesperson Problem(TSP)
    - VLSI layout
    - Robot navigation
    - Automatic assembly sequencing f. Internet searching

16. List the criteria to measure the performance of different search strategies.

- Completeness: Is the algorithm guaranteed to find a solution when there is one?
- Optimality: Does the strategy find the optimal solution?
- Time complexity: How long does it take to find a solution?
- Space complexity: How much memory is needed to perform the search?

17. Differentiate Uninformed Search (Blind search) and Informed Search (Heuristic Search) strategies.

Uninformed or Blind Search	Informed or Heuristic Search
• No additional information • beyond that provided in the problem definition • Not effective • No information about number of steps or path cost	• More effective • Uses problem-specific knowledge beyond • the definition of the problem itself.

18. Define Best-first-search.

Best-first search is an instance of the general TREE-SEARCH or GRAPH-SEARCH algorithm in which a node is selected for expansion based on the evaluation function f(n). Traditionally, the node with the lowest evaluation function is selected for expansion.

Chapter 2
PROBLEM SOLVING METHODS

2.1 Problem solving Methods

Problem solving – Introduction

Search is one of the operational tasks that characterize AI programs best. Almost every AI program depends on a search procedure to perform its prescribed functions. Problems are typically defined in terms of state, and solution corresponds to goal states.

Problem solving using search technique performs two sequences of steps:

1. Define the problem - Given problem is identified with its required initial and goal state.
2. Analyze the problem - The best search technique for the given: problem is chosen from different an AI search technique which derives one or more goal state in minimum number of states.

Types of Problems

In general the problem can be classified under anyone of the following four types which depends on two important properties. They are

1. Amount of knowledge, of the agent on the state and action description.
2. How the agent is connected to its environment through its percepts and actions?

The four different types of problems are:

- Single state problem
- Multiple state problem
- Contingency problem
- Exploration problem

Problem solving agents

Problem solving agent is one kind of goal based agent, where the agent decides what to do by finding sequence of actions that lead to desirable states. The complexity arises here is the knowledge about the formulation process, (from current state to outcome action) of the agent.

If the agent understood the definition of problem, it is relatively straight forward to construct a search process for finding solutions, which implies that problem solving agent should be an intelligent agent to maximize the performance measure.

The sequence of steps done by the intelligent agent to maximize the performance measure:

1. Goal formulation - based on current situation is the first step in problem solving. Actions that result to a failure case can be rejected without further consideration.
2. Problem formulation - is the process of deciding what actions and states to consider and follows goal formulation.
3. Search - is the process of finding different possible sequence of actions that lead to state of known value, and choosing the best one from the states.
4. Solution - a search algorithm takes a problem as input and returns a solution in the form of action sequence.
5. Execution phase - if the solution exists, the action it recommends can be carried out.

A simple problem solving agent

function SIMPLE-PROBLEM-SOLVING-AGENT(p)
returns an action
input : p, a percept
static: s, an action sequence, initially empty

state, some description of the current world state g, a goal initially null
problem, a problem formulation
state <- UPDATE-STATE(state, p)
if s is empty then
g <- FORMULATE-GOAL(state)
problem <-FORMULATE-PROBLEM(state,g)
s <- SEARCH(problem)
action <- RECOMMENDATION(s, state)
s <- REMAINDER(s, state)
return action

Note

RECOMMENDATION - first action in the sequence
REMAINDER - returns the rest
SEARCH - choosing the best one from the sequence of actions
FORMULATE-PROBLEM - sequence of actions and states that lead to goal state.
UPDATE-STATE - initial state is forced to next state to reach the goal state
Well-defined problems and solutions

A problem can be defined formally by four components:

a. initial state
b. successor function c. goal test
c. path cost

The initial state that the agent starts in Successor function (S) - Given a particular state x, S(x) returns a set of states reachable from x by any single action.

The goal test, which determines whether a given state is a goal state. Sometimes there is an explicit set of possible goal states, and the test simply checks whether the given state is one of them. A path cost function that assigns a numeric cost to each path. The problem- solving agent chooses a cost function that reflects its own performance measure.

A solution to a problem is a path from the initial state to a goal state

Operator - The set of possible actions available to the agent.

State space (or) state set space - The set of all possible states reachable from the initial state by any sequence of actions.

Path (state space) - The sequence of action leading from one state to another

The effectiveness of a search can be measured using three factors. They are:

1. Solution is identified or not?
2. Is it a good solution? If yes, then path cost to be minimum.
3. Search cost of the problem that is associated with time and memory required to find a solution.

For Example

Imagine an agent in the city of Arad, Romania, enjoying a touring holiday. Now, suppose the agent has a nonrefundable ticket to fly out of Bucharest the following day. In that case, it makes sense for the agent to adopt the goal of getting to Bucharest. The agent's task is to find out which sequence of actions will get it to a goal state.

2.2 Search Strategies

Search algorithms are one of the most important areas of Artificial Intelligence. This topic will explain all about the search algorithms in AI.

Problem-solving agents

In Artificial Intelligence, Search techniques are universal problem-solving methods. **Rational agents** or **Problem-solving agents** in AI mostly used these search strategies or algorithms to solve a specific problem and provide the best result. Problem-solving agents are the goal-based agents and use atomic representation. In this topic, we will learn various problem-solving search algorithms.

Search Algorithm Terminologies

Search: Searching is a step by step procedure to solve a search-problem in a given search space. A search problem can have three main factors:

1. **Search Space:** Search space represents a set of possible solutions, which a system may have.
2. **Start State:** It is a state from where agent begins **the search**.
3. **Goal test:** It is a function which observe the current state and returns whether the goal state is achieved or not.

- **Search tree:** A tree representation of search problem is called Search tree. The root of the search tree is the root node which is corresponding to the initial state.
- **Actions:** It gives the description of all the available actions to the agent.
- **Transition model:** A description of what each action do, can be represented as a transition model.
- **Path Cost:** It is a function which assigns a numeric cost
- to each path.
- **Solution:** It is an action sequence which leads from the start node to the goal node.
- **Optimal Solution:** If a solution has the lowest cost among all solutions.

Properties of Search Algorithms

Following are the four essential properties of search algorithms to compare the efficiency of these algorithms:

Completeness: A search algorithm is said to be complete if it guarantees to return a solution if at least any solution exists for any random input.

Optimality: If a solution found for an algorithm is guaranteed to be the best solution (lowest path cost) among all other solutions, then such a solution for is said to be an optimal solution.

Time Complexity: Time complexity is a measure of time for an algorithm to complete its task.

Space Complexity: It is the maximum storage space required at any point during the search, as the complexity of the problem.

Types of search algorithms

Based on the search problems we can classify the search algorithms into uninformed (Blind search) search and informed search (Heuristic search) algorithms.

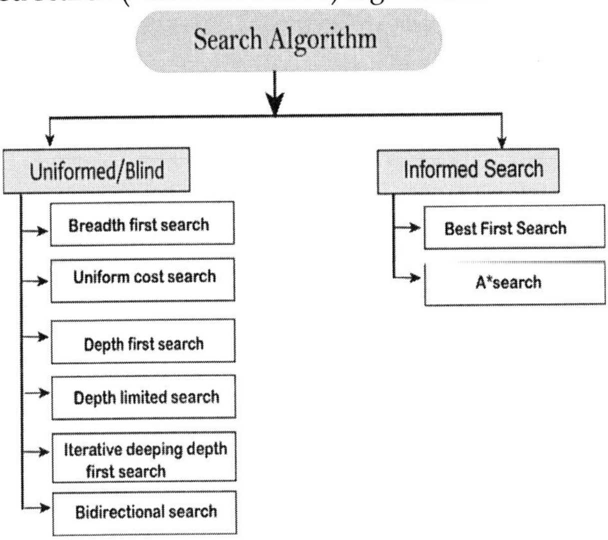

Uninformed/Blind Search

The uninformed search does not contain any domain knowledge such as closeness, the location of the goal. It operates in a brute-force way as it only includes information about how to traverse the tree and how to identify leaf and goal nodes. Uninformed search applies a way in which search tree is searched without any information about the search space like initial state operators and test for the goal, so it is also called blind search. It examines each node of the tree until it achieves the goal node.

It can be divided into five main types

- Breadth-first search
- Uniform cost search
- Depth-first search
- Iterative deepening depth-first search
- Bidirectional Search

Informed Search

Informed search algorithms use domain knowledge. In an informed search, problem information is available which can guide the search. Informed search strategies can find a solution more efficiently than an uninformed search strategy. Informed search is also called a Heuristic search.

A heuristic is a way which might not always be guaranteed for best solutions but guaranteed to find a good solution in reasonable time.

Informed search can solve much complex problem which could not be solved in another way.

An example of informed search algorithms is a traveling salesman problem.

1. Greedy Search
2. A* Search

2.3 Uninformed Search Algorithms

Uninformed search is a class of general-purpose search algorithms which operates in brute force-way. Uninformed search algorithms do not have additional information about state or search space other than how to traverse the tree, so it is also called blind search.

Following are the various types of uninformed search algorithms

1. Breadth-first Search
2. Depth-first Search
3. Depth-limited Search
4. Iterative deepening depth-first search
5. Uniform cost search
6. Bidirectional Search

1. Breadth-first Search

- Breadth-first search is the most common search strategy for traversing a tree or graph. This algorithm searches breadthwise in a tree or graph, so it is called breadth-first search.
- BFS algorithm starts searching from the root node of the tree and expands all successor node at the current level before moving to nodes of next level.
- The breadth-first search algorithm is an example of a general-graph search algorithm.

  Breadth-first search implemented using FIFO queue data structure.

Advantages

- BFS will provide a solution if any solution exists.
- If there are more than one solutions for a given problem, then BFS will provide the minimal solution which requires the least number of steps.

Disadvantages

- It requires lots of memory since each level of the tree must be saved into memory to expand the next level.
- BFS needs lots of time if the solution is far away from the root node.

Example

In the below tree structure, we have shown the traversing of the tree using BFS algorithm from the root node S to goal node K. BFS search algorithm traverse in layers, so it will follow the path which is shown by the dotted arrow, and the traversed path will be:

S---> A--->B---->C--->D---->G--->H--->E---->F---->I---->K

Breadth First Search

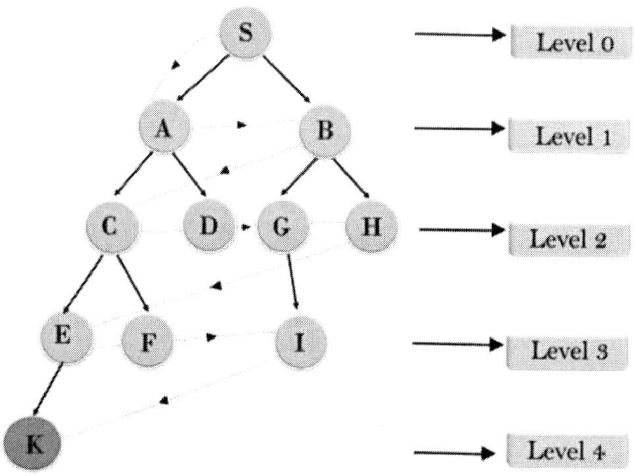

Time Complexity: Time Complexity of BFS algorithm can be obtained by the number of nodes traversed in BFS until the shallowest Node. Where the d= depth of shallowest solution and b is a node at every state.

T (b) = 1+b2+b3+.......+ bd= O (bd)

Space Complexity: Space complexity of BFS algorithm is given by the Memory size of frontier which is O(bd).

Completeness: BFS is complete, which means if the shallowest goal node is at some finite depth, then BFS will find a solution.

Optimality: BFS is optimal if path cost is a non-decreasing function of the depth of the node.

2. Depth-first Search

- Depth-first search is a recursive algorithm for traversing a tree or graph data structure.
- It is called the depth-first search because it starts from the root node and follows each path to its greatest depth node before moving to the next path.
- DFS uses a stack data structure for its implementation.
- The process of the DFS algorithm is similar to the BFS algorithm.

Advantage

- DFS requires very less memory as it only needs to store a stack of the nodes on the path from root node to the current node.
- It takes less time to reach to the goal node than BFS algorithm (if it traverses in the right path).

Disadvantage

- There is the possibility that many states keep re-occurring, and there is no guarantee of finding the solution.
- DFS algorithm goes for deep down searching and sometime it may go to the infinite loop.

Example

In the below search tree, we have shown the flow of depth- first search, and it will follow the order as:

Root node--->Left node ----> right node.

It will start searching from root node S, and traverse A, then B, then D and E, after traversing E, it will backtrack the tree as E has no other successor and still goal node is not found. After backtracking it will traverse node C and then G, and here it will terminate as it found goal node.

Depth First Search

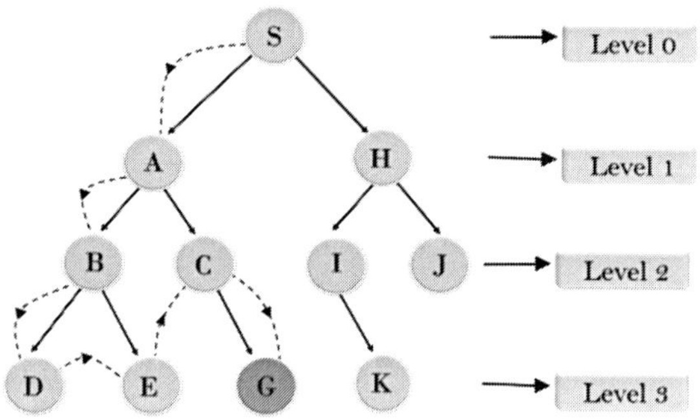

Completeness: DFS search algorithm is complete within finite state space as it will expand every node within a limited search tree.

Time Complexity: Time complexity of DFS will be equivalent to the node traversed by the algorithm. It is given
by:

T(n)= 1+ n2+ n3 +.........+ nm=O(nm)

Where, m= maximum depth of any node and this can be much larger than d (Shallowest solution depth)

Space Complexity: DFS algorithm needs to store only single path from the root node, hence space complexity of DFS is equivalent to the size of the fringe set, which is **O(bm)**.

Optimal: DFS search algorithm is non-optimal, as it may generate a large number of steps or high cost to reach to the goal node.

3. Depth-Limited Search Algorithm

A depth-limited search algorithm is similar to depth-first search with a predetermined limit. Depth-limited search can solve the drawback of the infinite path in the Depth-first search. In this algorithm, the node at the depth limit will treat as it has no successor nodes further.

Depth-limited search can be terminated with two Conditions of failure:

- Standard failure value: It indicates that problem does not have any solution.
- Cutoff failure value: It defines no solution for the problem within a given depth limit.

Advantages

Depth-limited search is Memory efficient.

Disadvantages

- Depth-limited search also has a disadvantage of incompleteness.
- It may not be optimal if the problem has more than one solution.

Depth Limited Search

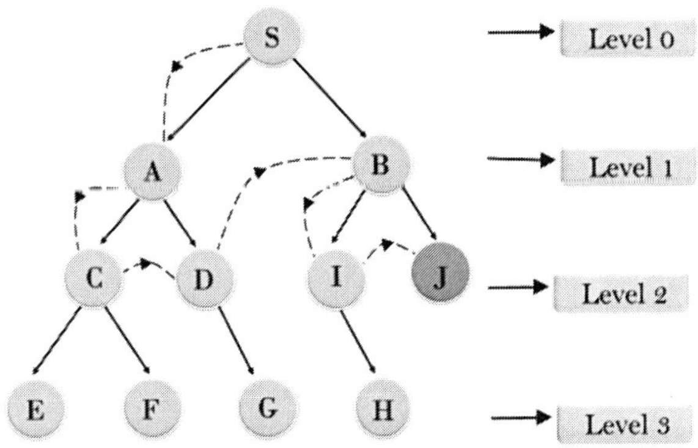

Example

Completeness: DLS search algorithm is complete if the solution is above the depth-limit.

Time Complexity: Time complexity of DLS algorithm is $O(b\ell)$.

Space Complexity: Space complexity of DLS algorithm is $O(b\times\ell)$.

Optimal: Depth-limited search can be viewed as a special case of DFS, and it is also not optimal even if $\ell>d$.

2.3. Uniform-cost Search Algorithm

Uniform-cost search is a searching algorithm used for traversing a weighted tree or graph. This algorithm comes into play when a different cost is available for each edge. The primary goal of the uniform-cost search is to find a path to the goal node which has the lowest cumulative cost. Uniform-cost search expands nodes according to their path costs form the root node. It can be used to solve any graph/tree where the optimal cost is in demand. A uniform-cost search algorithm is implemented by the priority queue. It gives maximum priority to the lowest cumulative cost. Uniform cost search is equivalent to BFS algorithm if the path cost of all edges is the same.

Advantages

- Uniform cost search is optimal because at every state the path with the least cost is chosen.

Disadvantages

- It does not care about the number of steps involve in searching and only concerned about path cost. Due to which this algorithm may be stuck in an infinite loop.

Example

Uniform Cost Search

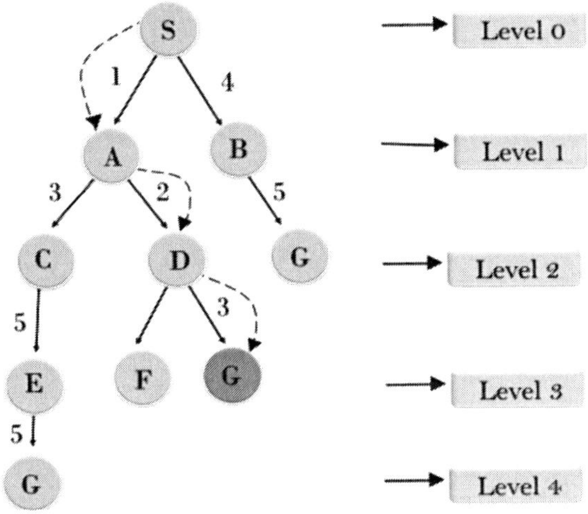

Completeness

Uniform-cost search is complete, such as if there is a solution, UCS will find it.

Time Complexity

Let C* is Cost of the optimal solution, and ε is each step to get closer to the goal node. Then the number of steps is = C*/ε+1. Here we have taken +1, as we start from state 0 and end to C*/ε.

Hence, the worst-case time complexity of Uniform-cost search isO(b1 + [C*/ε])/ .

Space Complexity

The same logic is for space complexity so, the worst-case space complexity of Uniform-cost search is O(b1 + [C*/ε]).

Optimal

Uniform-cost search is always optimal as it only selects a path with the lowest path cost.

2.4 Informed Search Algorithms

So far we have talked about the uninformed search algorithms which looked through search space for all possible solutions of the problem without having any additional knowledge about search space. But informed search algorithm contains an array of knowledge such as how far we are from the goal, path cost, how to reach to goal node, etc. This knowledge helps agents to explore less to the search space and find more efficiently the goal node.

The informed search algorithm is more useful for large search space. Informed search algorithm uses the idea of heuristic, so it is also called Heuristic search.

Heuristics function: Heuristic is a function which is used in Informed Search, and it finds the most promising path. It takes the current state of the agent as its input and produces the estimation of how close agent is from the goal. The heuristic method, however, might not always give the best solution, but it guaranteed to find a good solution in reasonable time. Heuristic function estimates how close a state is to the goal. It is represented by h(n), and it calculates the cost of an optimal path between the pair of states. The value of the heuristic function is always positive.

Admissibility of the heuristic function is given as h(n) <= h*(n)

Here h(n) is heuristic cost, and h*(n) is the estimated cost. Hence heuristic cost should be less than or equal to the estimated cost.

Pure Heuristic Search

Pure heuristic search is the simplest form of heuristic search algorithms. It expands nodes based on their heuristic value h(n). It maintains two lists, OPEN and CLOSED list. In the CLOSED list, it places those nodes which have already expanded and in the OPEN list, it places nodes which have yet not been expanded.

On each iteration, each node n with the lowest heuristic value is expanded and generates all its successors and n is placed to the closed list. The algorithm continues unit a goal state is found.

In the informed search we will discuss two main algorithms which are given below:

1. **Best First Search Algorithm(Greedy search)**
2. **A* Search Algorithm**

1. Best-first Search Algorithm (Greedy Search)

Greedy best-first search algorithm always selects the path which appears best at that moment. It is the combination of depth-first search and breadth-first search algorithms. It uses the heuristic function and search. Best-first search allows us to take the advantages of both algorithms. With the help of best-first search, at each step, we can choose the most promising node. In the best first search algorithm, we expand the node which is closest to the goal node and the closest cost is estimated by heuristic function, i.e.

f(n)= g(n).

Were, h(n)= estimated cost from node n to the goal.

The greedy best first algorithm is implemented by the priority queue.

Best first search algorithm

- **Step 1:** Place the starting node into the OPEN list.
- **Step 2:** If the OPEN list is empty, Stop and return failure.
- **Step 3:** Remove the node n, from the OPEN list which has the lowest value of h(n), and places it in the CLOSED list.
- **Step 4:** Expand the node n, and generate the successors of node n.
- **Step 5:** Check each successor of node n, and find whether

any node is a goal node or not. If any successor node is goal node, then return success and terminate the search, else proceed to Step 6.

- **Step 6:** For each successor node, algorithm checks for evaluation function f(n), and then check if the node has been in either OPEN or CLOSED list. If the node has not been in both list, then add it to the OPEN list.
- **Step 7:** Return to Step 2.

Advantages

- Best first search can switch between BFS and DFS by gaining the advantages of both the algorithms.
- This algorithm is more efficient than BFS and DFS algorithms.

Disadvantages

- It can behave as an unguided depth-first search in the worst case scenario.
- It can get stuck in a loop as DFS.
- This algorithm is not optimal.

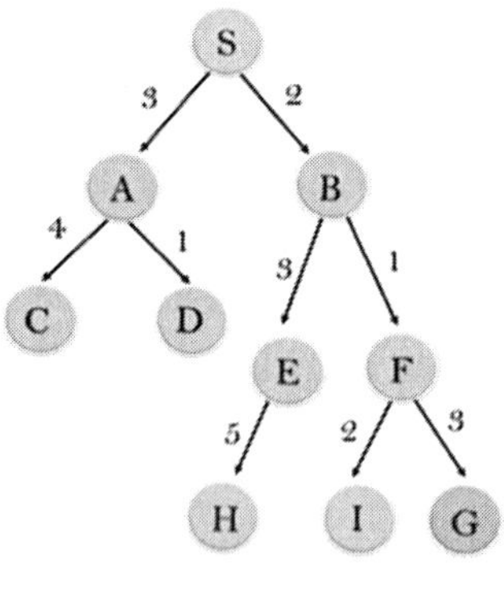

node	H (n)
A	12
B	4
C	7
D	3
E	8
F	2
H	4
I	9
S	13
G	0

Consider the below search problem, and we will traverse it using greedy best-first search. At each iteration, each node is expanded using evaluation function f(n)=h(n) , which is given in the below table.

In this search example, we are using two lists which are **OPEN** and **CLOSED** Lists. Following are the iteration for traversing the above example.

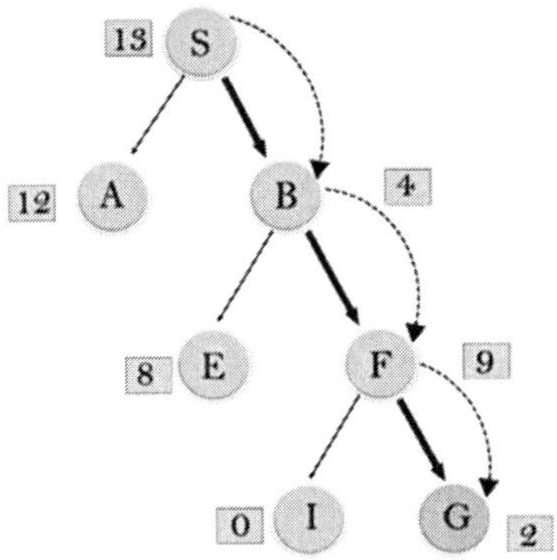

Expand the nodes of S and put in the CLOSED list
Initialization: Open [A, B], Closed [S] **Iteration 1:** Open [A], Closed [S, B] **Iteration 2:** Open [E, F, A], Closed [S, B]
: Open [E, A], Closed [S, B, F]
Iteration 3: Open [I, G, E, A], Closed [S, B, F]
: Open [I, E, A], Closed [S, B, F, G]
Hence the final solution path will be: **S----> B----->F----> G**
Time Complexity: The worst case time complexity of Greedy best first search is O(bm).
Space Complexity: The worst case space complexity of Greedy best first search is O(bm). Where, m is the maximum depth of the search space.
Complete: Greedy best-first search is also incomplete, even if the given state space is finite.
Optimal: Greedy best first search algorithm is not optimal.

2. A* Search Algorithm

A* search is the most commonly known form of best-first search. It uses heuristic function h(n), and cost to reach the node n from the start state g(n). It has combined features of UCS and greedy best-first search, by which it solve the problem efficiently. A* search algorithm finds the shortest path through the search space using the heuristic function. This search algorithm expands less search tree and provides optimal result faster. A* algorithm is similar to UCS except that it uses g(n)+h(n) instead of g(n).

In A* search algorithm, we use search heuristic as well as the cost to reach the node. Hence we can combine both costs as following, and this sum is called as a fitness number.

$$f(n) = g(n) + h(n)$$

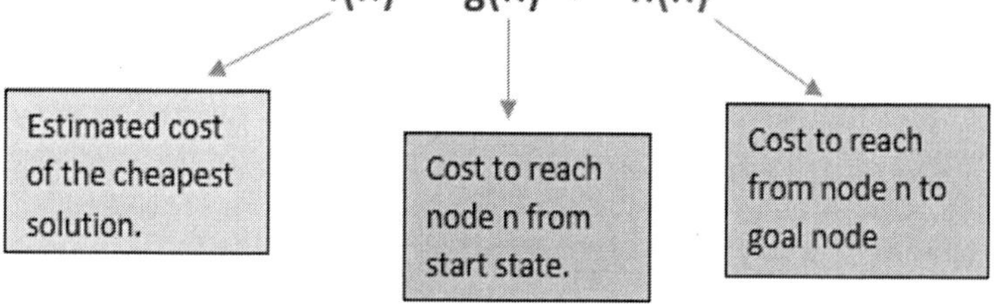

Algorithm of A* search

Step1: Place the starting node in the OPEN list.

Step 2: Check if the OPEN list is empty or not, if the list is empty then return failure and stops.

Step 3: Select the node from the OPEN list which has the smallest value of evaluation function (g+h), if node n is goal node then return success and stop, otherwise

Step 4: Expand node n and generate all of its successors, and put n into the closed list. For each successor n', check whether n' is already in the OPEN or CLOSED list, if not then compute evaluation function for n' and place into Open list.

Step 5: Else if node n' is already in OPEN and CLOSED, then it should be attached to the back pointer which reflects the lowest g(n') value.

Step 6: Return to **Step 2.**

Advantages

- A* search algorithm is the best algorithm than other search algorithms.
- A* search algorithm is optimal and complete.
- This algorithm can solve very complex problems.

Disadvantages

- It does not always produce the shortest path as it mostly based on heuristics and approximation.
- A* search algorithm has some complexity issues.
- The main drawback of A* is memory requirement as it keeps all generated nodes in the memory, so it is not practical for various large-scale problems.

Example

In this example, we will traverse the given graph using the A* algorithm. The heuristic value of all states is given in the below table so we will calculate the f(n) of each state using the formula f(n)= g(n) + h(n), where g(n) is the cost to reach any node from start state.

Here we will use OPEN and CLOSED list.

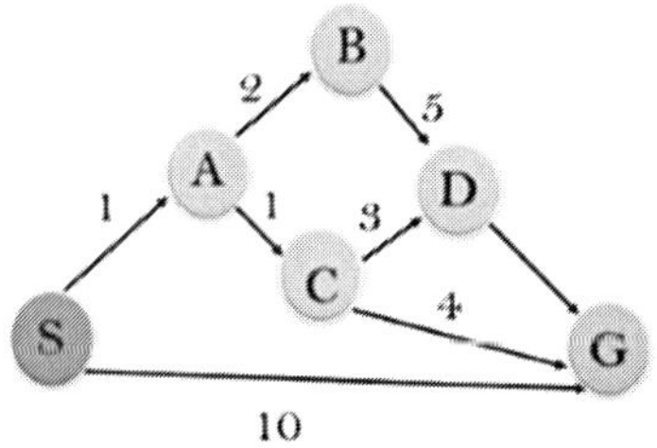

State	h(n)
S	5
A	3
B	4
C	2
D	6
G	0

Solution:

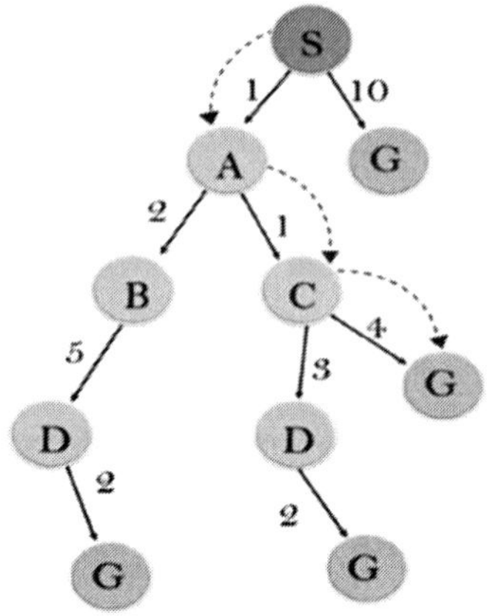

Initialization: {(S, 5)}
Iteration1: {(S--> A, 4), (S-->G, 10)}
Iteration2: {(S--> A-->C, 4), (S--> A-->B, 7), (S-->G, 10)}
Iteration3: {(S--> A-->C--->G, 6), (S--> A-->C--->D, 11), (S--> A-->B, 7), (S-->G, 10)}
Iteration 4 will give the final result, as S--->A--->C--->G it provides the optimal path with cost 6.

Points to remember

- A* algorithm returns the path which occurred first, and it does not search for all remaining paths.
- The efficiency of A* algorithm depends on the quality of heuristic.
- A* algorithm expands all nodes which satisfy the condition f(n)<="" li="">

Complete: A* algorithm is complete as long as:

- Branching factor is finite.
- Cost at every action is fixed.

Optimal: A* search algorithm is optimal if it follows below two conditions:

- Admissible: the first condition requires for optimality is that h(n) should be an admissible heuristic for A* tree search. An admissible heuristic is optimistic in nature.

- Consistency: Second required condition is consistency for only A* graph-search.

If the heuristic function is admissible, then A* tree search will always find the least cost path.

Time Complexity: The time complexity of A* search algorithm depends on heuristic function, and the number of nodes expanded is exponential to the depth of solution d. So the time complexity is O(b^d), where b is the branching factor.

Space Complexity: The space complexity of A* search algorithm is O(b^d)

2.5 Local Search Algorithms and Optimization Problems

Hill Climbing Algorithm in Artificial Intelligence

- Hill climbing algorithm is a local search algorithm which continuously moves in the direction of increasing elevation/value to find the peak of the mountain or best solution to the problem. It terminates when it reaches a peak value where no neighbor has a higher value.
- Hill climbing algorithm is a technique which is used for optimizing the mathematical problems. One of the widely discussed examples of Hill climbing algorithm is Traveling-salesman Problem in which we need to minimize the distance traveled by the salesman.
- It is also called greedy local search as it only looks to its good immediate neighbor state and not beyond that.
- A node of hill climbing algorithm has two components which are state and value.
- Hill Climbing is mostly used when a good heuristic is available.
- In this algorithm, we don't need to maintain and handle the search tree or graph as it only keeps a single current

state.

- Features of Hill Climbing:

Following are some main features of Hill Climbing Algorithm:

- **Generate and Test variant:** Hill Climbing is the variant of Generate and Test method. The Generate and Test method produce feedback which helps to decide which direction to move in the search space.
- **Greedy approach:** Hill-climbing algorithm search moves in the direction which optimizes the cost.
- **No backtracking:** It does not backtrack the search space, as it does not remember the previous states.

State-space Diagram for Hill Climbing

The state-space landscape is a graphical representation of the hill-climbing algorithm which is showing a graph between various states of algorithm and Objective function/Cost.

On Y-axis we have taken the function which can be an objective function or cost function, and state-space on the x-axis. If the function on Y-axis is cost then, the goal of search is to find the global minimum and local minimum. If the function of Y-axis is Objective function, then the goal of the search is to find the global maximum and local maximum.

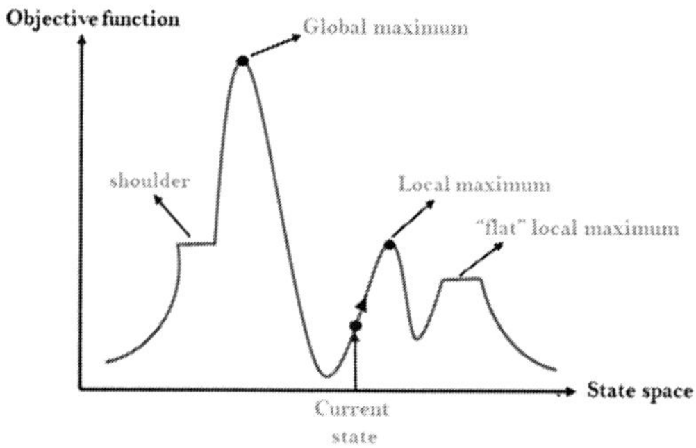

Different regions in the state space landscape

Local Maximum: Local maximum is a state which is better than its neighbor states, but there is also another state which is higher than it.

Global Maximum: Global maximum is the best possible state of state space landscape. It has the highest value of objective function.

Current state: It is a state in a landscape diagram where an agent is currently present.

Flat local maximum: It is a flat space in the landscape where all the neighbor states of current states have the same value.

Shoulder: It is a plateau region which has an uphill edge.

Types of Hill Climbing Algorithm

1. Simple hill Climbing
2. Steepest-Ascent hill-climbing
3. Stochastic hill Climbing

1. Simple Hill Climbing

Simple hill climbing is the simplest way to implement a hill climbing algorithm. **It only evaluates the neighbor node state at a time and selects the first one which optimizes current cost and set it as a current state**. It only checks it's one successor state, and if it finds better than the current state, then move else be in the same state. This algorithm has the following features:

- Less time consuming
- Less optimal solution and the solution is not guaranteed

Algorithm for Simple Hill Climbing:

- **Step 1:** Evaluate the initial state, if it is goal state then return success and Stop.

- **Step 2:** Loop Until a solution is found or there is no new operator left to apply.
- **Step 3:** Select and apply an operator to the current state.
- **Step 4:** Check new state:
    1. If it is goal state, then return success and quit.
    2. Else if it is better than the current state then assign new state as a current state.
    3. Else if not better than the current state, then return to step2.
- **Step 5:** Exit.

2. Steepest-Ascent hill climbing

The steepest-Ascent algorithm is a variation of simple hill climbing algorithm. This algorithm examines all the neighboring nodes of the current state and selects one neighbor node which is closest to the goal state. This algorithm consumes more time as it searches for multiple neighbors

Algorithm for Steepest-Ascent hill climbing

- **Step 1:** Evaluate the initial state, if it is goal state then return success and stop, else make current state as initial state.
- **Step 2:** Loop until a solution is found or the current state does not change.

1. Let SUCC be a state such that any successor of the current state will be better than it.
2. For each operator that applies to the current state:
    a. Apply the new operator and generate a new state.
    b. Evaluate the new state.
    c. If it is goal state, then return it and quit, else compare it to the SUCC.

d. If it is better than SUCC, then set new state as SUCC.
e. If the SUCC is better than the current state, then set current state to SUCC.

- **Step 5:** Exit.

3. Stochastic hill climbing

Stochastic hill climbing does not examine for all its neighbor before moving. Rather, this search algorithm selects one neighbor node at random and decides whether to choose it as a current state or examine another state.

Problems in Hill Climbing Algorithm

1. Local Maximum

A local maximum is a peak state in the landscape which is better than each of its neighboring states, but there is another state also present which is higher than the local maximum.

Solution: Backtracking technique can be a solution of the local maximum in state space landscape. Create a list of the promising path so that the algorithm can backtrack the search space and explore other paths as well.

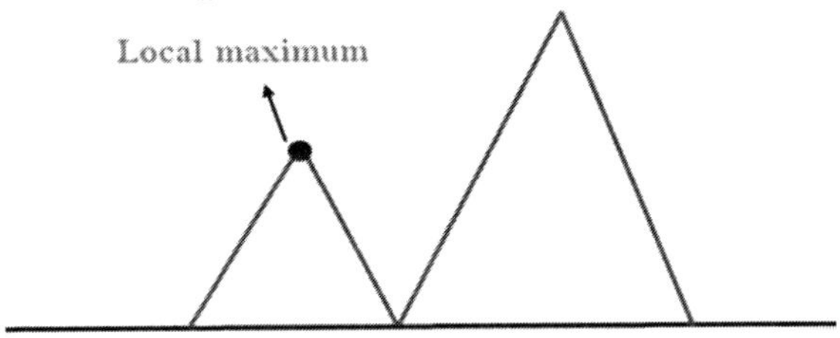

2. Plateau

A plateau is the flat area of the search space in which all the neighbor states of the current state contains the same value, because of this algorithm does not find any best direction to move. A hill-climbing search might be lost in the plateau area.

Solution: The solution for the plateau is to take big steps or very little steps while searching, to solve the problem. Randomly select a state which is far away from the current state so it is possible that the algorithm could find non-plateau region.

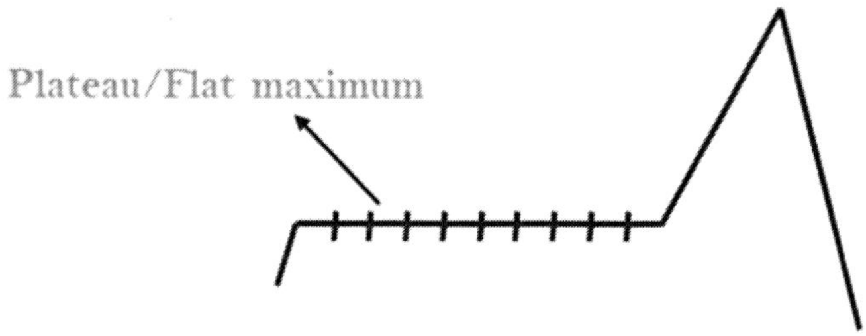

3. Ridges

A ridge is a special form of the local maximum. It has an area which is higher than its surrounding areas, but itself has a slope, and cannot be reached in a single move.

Solution: With the use of bidirectional search, or by moving in different directions, we can improve this problem.

Simulated Annealing

A hill-climbing algorithm which never makes a move towards a lower value guaranteed to be incomplete because it can get stuck on a local maximum. And if algorithm applies a random walk, by moving a successor, then it may complete but not efficient. **Simulated Annealing** is an algorithm which yields both efficiency and completeness.

In mechanical term **Annealing** is a process of hardening a metal or glass to a high temperature then cooling gradually, so this allows the metal to reach a low-energy crystalline state.

The same process is used in simulated annealing in which the algorithm picks a random move, instead of picking the best move. If the random move improves the state, then it follows the same path. Otherwise, the algorithm follows the path which has a probability of less than 1 or it moves downhill and chooses another path.

2.6 Constraint Satisfaction Problems (CSP)

Constraint satisfaction problems (CSP) are mathematical problems where one must find states or objects that satisfy a number of constraints or criteria. A constraint is a restriction of the feasible solutions in an optimization problem.

Some examples for CSP's are:

1. The n-queens problem
2. A crossword puzzle
3. A map coloring problem
4. The Boolean satisfiability problem
5. A crypt arithmetic problem

All these examples and other real life problems like time table scheduling, transport scheduling, floor planning etc. are instances of the same pattern, A Constraint Satisfaction Problem(or CSP) is defined by a set of variables {X1 ,X2 ,....Xn,} and a set of constraints {C1, C2 ,...,Cm }.

Each variable Xi has a nonempty domain D, of possible values. Each constraint Ci involves some subset of variables and specifies the allowable combinations of values for that subset.

A State of the problem is defined by an assignment of values to some or all of the variables,{Xi = vi , Xj = vj ,...}. An assignment that does not violate any constraints is called a consistent or legal assignment.

A State of the problem is defined by an assignment of values to some or all of the variables,{Xi = vi , Xj = vj ,…}. An assignment that does not violate any constraints is called a consistent or legal assignment.

Example for Constraint Satisfaction Problem The map coloring problem. The task of coloring each region red, green or blue in such a way that no neighboring regions have the same color.

2.7 Constraint propagation

Although forward checking detects many inconsistencies, it does not detect all of them.

Constraint propagation is the general term for propagating the implications of a constraint on one variable onto other variables.

Constraint propagation repeatedly enforces constraints locally to detect inconsistencies. This propagation can be done with different types of consistency techniques. They are:

- Node consistency (one consistency)
- Arc consistency (two consistencies)
- Path consistency (K-consistency)

Arc Consistency

Simplest consistency technique

The node representing a variable V in constraint graph is node consistent if for every value X in the current domain of V, each unary constraint on V is satisfied.

The node inconsistency can be eliminated by simply removing those values from the domain D of each variable V that do not satisfy unary constraint on V.

Arc Consistency

The idea of arc consistency provides a fast method of constraint propagation that is substantially stronger than forward checking. Here, 'arc' refers to a directed arc in the constraint graph, such as the arc from SA to NSW. Given the current domains of SA and NSW, the arc is consistent if, for every value x of SA, there is some value y of NSW that is consistent with x.

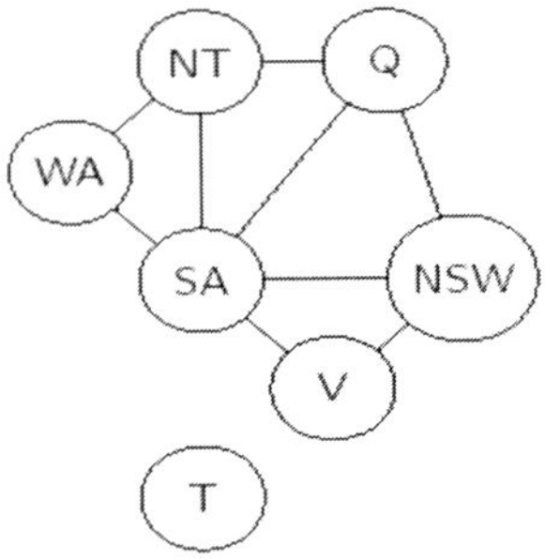

Figure: Australian Territories

In the constraint graph, binary constraint corresponds to arc. Therefore this type of consistency is called arc consistency.

Arc (vi , vj) is arc consistent if for every value X the current domain of vi there is some value Y in the domain of vj such vi =X and vj =Y is permitted by the binary constraint between vi and vj

Arc-consistency is directional ie if an arc (vi, vj) is consistent than it does not automatically mean that (vj, vi) is also consistent.

An arc (vi, vj) can be made consistent by simply deleting those values from the domain of Di for which there is no corresponding value in the domain of Dj such that the binary constraint between Vi and vj is satisfied - It is an earlier detection of inconstency that is not detected by forward checking method.

The different versions of Arc consistency algorithms are exist such as AC-I, AC2,AC-3, AC-4, AC-S; AC-6 & AC-7, but frequently used are AC-3 or AC-4.

AC - 3 Algorithm

In this algorithm, queue is used to cheek the inconsistent arcs. When the queue is not empty do the following steps:

Remove the first arc from the queue and check for consistency.

- If it is inconsistent remove the variable from the domain and add a new arc to the queue

Repeat the same process until queue is empty

```
function AC-3( csp) returns the CSP, possibly with reduced
domains
inputs: csp, a binary CSP with variables {X1, X2, . . . ,
Xn}
local variables: queue, a queue of arcs, initially all the arcs in
csp
while queue is not empty do
(Xi, Xj) <- REMOVE-FIRST(queue)
if REMOVE-INCONSISTENT-VALUEXS(xi,xj) then for each
Xk in NEIGHBORS[Xj) do
add (Xk , Xi)to queue
```

```
function REMOVE-INCONSISTENT-VALUEXS(xi,xj) returns
true iff we remove a value
removed <-false
for each x in DOMAIN[xi] do
if no value y in DOMAIN[xj] allows (x, y) to satisfy the
constraint between Xi and Xj
then delete x from DOMAIN[Xi]removed <- true
return removed
```

k-Consistency (path Consistency)

A CSP is k-consistent if, for any set of k - 1 variables and for any consistent assignment to those variables, a consistent value can always be assigned to any kth variable

a. consistency means that each individual variable by itself is consistent; this is also called node consistency.

b. consistency is the same as arc consistency.

c. consistency means that any pair of adjacent variables can always be extended to a third neighboring variable; this is also called **path consistency**

2.8 Backtracking search

Backtracking search, a form of depth-first search, is commonly used for solving CSPs. Inference can be interwoven with search.

Commutatively: CSPs are all commutative. A problem is commutative if the order of application of any given set of actions has no effect on the outcome.

Backtracking search: A depth-first search that chooses values for one variable at a time and backtracks when a variable has no legal values left to assign.

Backtracking algorithm repeatedly chooses an unassigned variable, and then tries all values in the domain of that variable in turn, trying to find a solution. If an inconsistency is detected, then BACKTRACK returns failure, causing the previous call to try another value.

There is no need to supply BACKTRACKING-SEARCH with a domain-specific initial state, action function, transition model, or goal test.

BACKTRACKING-SARCH keeps only a single representation of a state and alters that representation rather than creating a new ones.

```
function BACKTRACKING-SEARCH(csp) returns a solution, or failure
  return BACKTRACK({ }, csp)

function BACKTRACK(assignment, csp) returns a solution, or failure
  if assignment is complete then return assignment
  var ← SELECT-UNASSIGNED-VARIABLE(csp)
  for each value in ORDER-DOMAIN-VALUES(var, assignment, csp) do
     if value is consistent with assignment then
        add {var = value} to assignment
        inferences ← INFERENCE(csp, var, value)
        if inferences ≠ failure then
          add inferences to assignment
          result ← BACKTRACK(assignment, csp)
          if result ≠ failure then
             return result
     remove {var = value} and inferences from assignment
  return failure
```

2.9 Game playing

- Mini-max algorithm is a recursive or backtracking algorithm which is used in decision-making and game theory. It provides an optimal move for the player assuming that opponent is also playing optimally.
- Mini-Max algorithm uses recursion to search through the game-tree.
- Min-Max algorithm is mostly used for game playing in AI. Such as Chess, Checkers, tic-tac-toe, go, and various tow-players game. This Algorithm computes the minimax decision for the current state.
- In this algorithm two players play the game, one is called MAX and other is called MIN.
- Both the players fight it as the opponent player gets the minimum benefit while they get the maximum benefit.
- Both Players of the game are opponent of each other, where MAX will select the maximized value and MIN will select the minimized value.
- The minimax algorithm performs a depth-first search algorithm for the exploration of the complete game tree.

- The minimax algorithm proceeds all the way down to the terminal node of the tree, then backtrack the tree as the recursion.

Pseudo-code for MinMax Algorithm

```
function minimax(node, depth, maximizingPlayer) is
if depth ==0 or node is a terminal node then
return static evaluation of node
if MaximizingPlayer then // for Maximizer Player
maxEva= -infinity
for each child of node do
eva= minimax(child, depth-1, false)
maxEva= max(maxEva,eva)//gives Maximum of the values
return maxEva
else // for Minimizer player
inEva= +infinity
for each child of node do
eva= minimax(child, depth-1, true)
minEva= min(minEva, eva)//gives minimum of the values
return minEva
```

Initial call

Minimax(node, 3, true)

Working of Min-Max Algorithm

- The working of the minimax algorithm can be easily described using an example. Below we have taken an example of game-tree which is representing the two- player game.
- In this example, there are two players one is called Maximizer and other is called Minimizer.
- Maximizer will try to get the Maximum possible

score, and Minimizer will try to get the minimum possible score.

- This algorithm applies DFS, so in this game-tree, we have to go all the way through the leaves to reach the terminal nodes.
- At the terminal node, the terminal values are given so we will compare those value and backtrack the tree until the initial state occurs.

Following are the main steps involved in solving the two-player game tree:

Step-1: In the first step, the algorithm generates the entire game-tree and apply the utility function to get the utility values for the terminal states. In the below tree diagram, let's take A is the initial state of the tree. Suppose maximizer takes first turn which has worst-case initial value =- infinity, and minimizer will take next turn which has worst-case initial value = +infinity

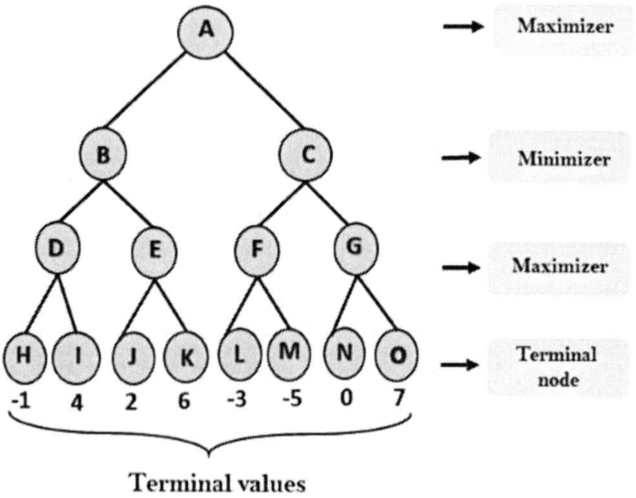

Step 2: Now, first we find the utilities value for the Maximizer, its initial value is -∞, so we will compare each

value in terminal state with initial value of Maximizer and determines the higher nodes values. It will find the maximum among the all.

- For node D max(-1,- -∞) => max(-1,4)= 4
- For Node E max(2, -∞) => max(2, 6)= 6
- For Node F max(-3, -∞) => max(-3,-5) = -3
- For node G max(0, -∞) = max(0, 7) = 7

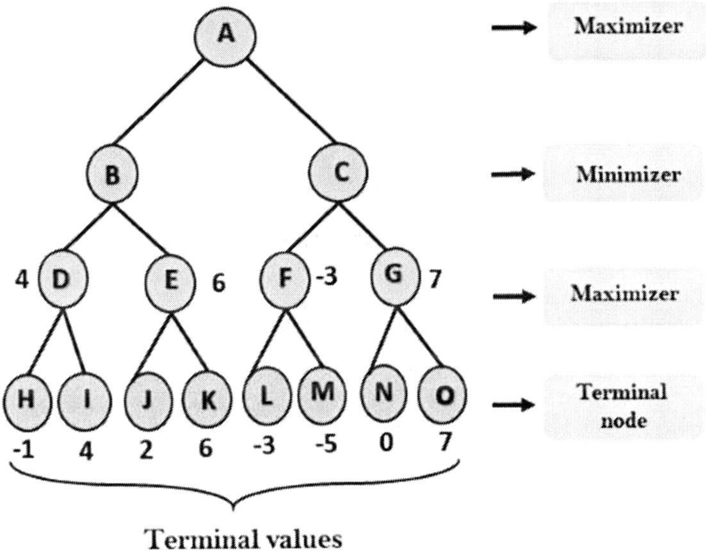

Step 3: In the next step, it's a turn for minimizer, so it will compare all nodes value with +∞, and will find the 3rd layer node values.

- For node B= min(4,6) = 4
- For node C= min (-3, 7) = -3

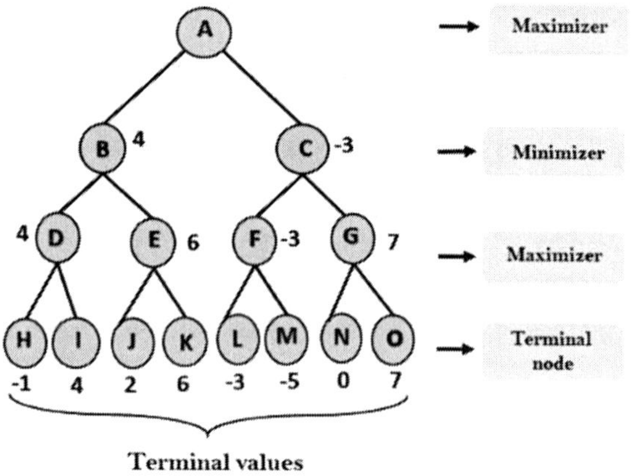

Step 3: Now it's a turn for Maximizer, and it will again choose the maximum of all nodes value and find the maximum value for the root node. In this game tree, there are only 4 layers, hence we reach immediately to the root node, but in real games, there will be more than 4 layers.

- For node A max(4, -3)= 4

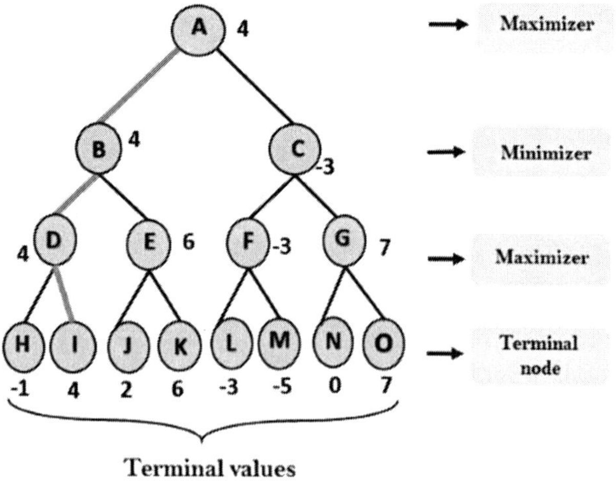

That was the complete workflow of the minimax two player game.

Properties of Mini-Max algorithm

- **Complete-** Min-Max algorithm is Complete. It will definitely find a solution (if exist), in the finite search tree.
- **Optimal-** Min-Max algorithm is optimal if both opponents are playing optimally.
- **Time complexity-** As it performs DFS for the game-tree, so the time complexity of Min-Max algorithm is $\mathbf{O(b^m)}$, where b is branching factor of the game-tree, and m is the maximum depth of the tree.
- **Space Complexity-** Space complexity of Mini-max algorithm is also similar to DFS which is **O(bm)**.

Limitation of the min-max Algorithm

The main drawback of the min-max algorithm is that it gets really slow for complex games such as Chess, go, etc. This type of games has a huge branching factor, and the player has lots of choices to decide. This limitation of the min-max algorithm can be improved from **alpha-beta pruning** which we have discussed in the next topic.

2.10 Alpha - Beta Pruning

Pruning - The process of eliminating a branch of the search tree from consideration without examining is called pruning.

The two parameters of pruning technique are

Alpha (α): Best choice for the value of MAX along the path (or) lower bound on the value that on maximizing node may be ultimately assigned.

Beta (β): Best choice for the value of MIN along the path (or) upper bound on the value that a minimizing node may be ultimately assigned.

Alpha - Beta pruning : The α and β values are applied to a

minimax tree, it returns the same move as minimax, but prunes away branches that cannot possibly influence the final decision is called Alpha - Beta pruning (or) Cutoff Consider again the two-ply game tree from

Let the two unevaluated successors of node C have values x and y and let z be the minimum of x and y. The value of the root node is given by

MINIMAX-VALUE(ROOT)=max((min(3,12,8),min(2,x,y),min(14,5,2))

= *max(3, min(2, x,* y), *2)*

= *max(3, z, 2)* where z <=2

= **3.**

In other words, the value of the root and hence the minimax decision are *independent* of the values of the pruned leaves x and y.

Stages in the calculation of the optimal decision for the game tree

a. The first leaf below B has the value 3. Hence, B, which is a MIN node, has a value of *at most* 3

b. The second leaf below B has a value of 12; MIN would avoid this move, so the value of B is still at most 3

c. The third leaf below B has a value of 8; we have seen all B's successors, so the value of B is exactly 3. Now, we can infer that the value of the root is at least 3, because MAX has a choice worth 3 at the root.

d. The first leaf below C has the value 2. Hence, C, which is a MIN node, has a value of at most 2. But we know that B is worth 3, so MAX would never choose C. Therefore, there is no point in looking at the other successors of C. This is an example of alpha-beta pruning.

e. The first leaf below D has the value 14, so D is worth atmost 14. This is still higher than MAX'S best alternative (i.e., 3), so we need to keep exploring D's successors. Notice also that we now have bounds on all of the successors of the root, so the root's value is also at most 14.

f. The second successor of D is worth 5, so again we need to keep exploring. The third successor is worth 2, so now D is worth exactly 2. MAX'S decision at the root is to move to B, giving a value of 3.

The alpha-beta search algorithm

function ALPHA-BETA-SEARCH (returns an) action

inputs: state, current state in game

v<- M A X -V A L U E (State,- α , +α)

return the action in SUCCESSORS(state) with value v

function MIN - VALUE (state, α , β) **returns** α utility value

inputs: state, current state in game

α , the value of the best alternative for MAX along the path to state

β , the value of the best alternative for MIN along the path to state

if TERMINAL-TEST(state) then return) UTILITY(s t a t e)

v<- +∞

for α , S in S U C C E S S O R S (state) do

v <- MIN(v,MAX-VALUE(S, α , β))

if v <= α **then return** v

β <- MIN(β , v)

return v

Effectiveness of Alpha – Beta Pruning

Alpha - Beta pruning algorithm needs to examine only O(bd/2) nodes to pick the best move, instead of O(bd) with minimax algorithm, that is effective branching factor is β instead of b

Imperfect Real Time Decisions

The min-max algorithm generates the entire game search space, whereas the alpha-beta algorithm allows us to prune large parts of it. However, alpha-beta still has to search all the way to terminal states for at least a portion of the search space. This depth is usually not practical, because moves must be made in a reasonable amount of time-typically a few minutes at most.

Shannon's proposed instead that programs should cut off the search earlier and apply a heuristic evaluation function to states in the search, effectively turning non terminal nodes into terminal leaves

□ The utility function is replaced by an Evaluation function

□ The terminal test is replaced by a Cut-off test

2-QUESTION BANK

Part - A

1. What is informed search?

One that uses problem - specific knowledge beyond the definition of the problem itself and it can find solutions more efficiently than an uninformed strategy.

2. What is the use of QUEUING_FN?

QUEUING_FN inserts asset of elements into the queue. Different varieties of queuing fn produce different varieties of the search algorithm.

3. Mention the criteria for the evaluation of search strategy.

There are 4 criteria: Completeness, time complexity, space complexity, optimality

4. List the various search strategies. a. BFS

a. BFS
b. Uniform cost search c. DFS
d. Depth limited search
e. Iterative deepening search f. Bidirectional search

5. List the various informed search strategy. Best first search–greedy search, A* search

Memory bounded search-Iterative deepening A*search - simplified memory bounded

A*search -Iterative improvement search –hill climbing - simulated annealing

6. What is Best First Search?

Best First Search is an instance of the general TREE SEARCH or GRAPH SEARCH algorithm in which a node is selected for expansion based on an evaluation function, f(n).

7. Define Evaluation function, f(n).

A node with the lowest evaluation is selected for expansion, because evaluation measures distance to the goal.

8. Define Heuristic function, h (n).

h (n) is defined as the estimated cost of the cheapest path from node n to a goal node.

9. Define Greedy Best First Search.

It expands the node that is closest to the goal (i.e.) to reach solution in a quicker way. It is done by using the heuristic function: f(n) = h(n).

10. Define A* search.

A* search evaluates nodes by combining g(n), the cost to reach the node and h(n), the cost to get from the node to the goal.

The sum of the horizontal and vertical distances of the tiles from their goal positions in a 15 puzzle problem is called Manhattan distance (or) city block distance.

11. Give the drawback of DFS.

The drawback of DFS is that it can get stuck going down the wrong path. Many problems have very deep or even infinite search tree. So dfs will never be able to recover from an unlucky choice at one of the nodes near the top of the tree. So DFS should be avoided for search trees with large or infinite maximum depths

12. Define Branching factor b*.

Uniform tree of depth d would have to be in order to contain N+1 nodes is called branching factor.

13. Write the time & space complexity associated with depth limited search. ime complexity =O (bd), b-branching factor, d-depth of tree Space complexity=o (bl)

14. What is Released problems?

A problem with fewer restrictions on the actions is called a relaxed problem.

15. What is a pattern database?

This database is the storage of exact solution costs for every possible sub problem instance.

16. What is a disjoint pattern database?

The sum of the two costs is still a lower bound on the cost of solving the entire problem is called a disjoint pattern database.

17. What is local search?

It operates using a single current state rather than multiple paths and generally moves only to neighbors of that state.

18. Define Optimization Problems.

The aim of this problem is to find the best state according to an objective function.

19. What are the 2 parts of Landscape?

1. Location defined by the state.
2. Elevation defined by the value of the heuristic cost function (or) objective function.

Chapter 3
KNOWLEDGE REPRESENTATION

3.1 First Order Predicate Logic

In the topic of Propositional logic, we have seen that how to represent statements using propositional logic. But unfortunately, in propositional logic, we can only represent the facts, which are either true or false. PL is not sufficient to represent the complex sentences or natural language statements. The propositional logic has very limited expressive power. Consider the following sentence, which we cannot represent using PL logic.

- **"Some humans are intelligent", or**
- **"Sachin likes cricket."**

To represent the above statements, PL logic is not sufficient, so we required some more powerful logic, such as first-order logic.

First-Order logic

- First-order logic is another way of knowledge representation in artificial intelligence. It is an extension to propositional logic.
- FOL is sufficiently expressive to represent the natural language statements in a concise way.
- First-order logic is also known as **Predicate logic or First-order predicate logic**. First-order logic is a powerful language that develops information about the objects in a more **easy way and can also express the relationship between those objects.**
- **First-order logic (like natural language) does not only assume that the world contains facts like propositional logic but also assumes the following things in the world:**

- Objects: **A, B, people, numbers, colors, wars, theories, squares, pits, wumpus, ...**
- Relations: It can be unary relation such as: **red, round, is adjacent,** or n-any relation such as: **the sister of, brother of, has color, comes between**
- Function: **Father of, best friend, third inning of, end of,** ......
- **As a natural language, first-order logic also has two main parts:**
- Syntax
- Semantics

Syntax of First-Order logic

- The syntax of FOL determines which collection of symbols is a logical expression in first-order logic. The basic syntactic elements of first-order logic are symbols. We write statements in short-hand notation in FOL.
- Basic Elements of First-order logic:
- Following are the basic elements of FOL syntax:

Constant	**1, 2, A, John, Mumbai, cat,....**
Variables	**x, y, z, a, b,....**
Predicates	**Brother, Father, >,....**
Function	**sqrt, LeftLegOf, ...**
Connectives	$\land$, $\lor$, $\neg$, $\Rightarrow$, $\Leftrightarrow$
Equality	==
Quantifier	$\forall$, $\exists$

Atomic sentences

- Atomic sentences are the most basic sentences of first-order logic. These sentences are formed from a predicate symbol followed by a parenthesis with a sequence of terms.

- **We can represent atomic sentences as** Predicate (term1, term2, ..., term n).
- Example: Ravi and Ajay are brothers: => Brothers (Ravi, Ajay). Chinky is a cat: => cat (Chinky).

Complex Sentences

- **Complex sentences are made by combining atomic sentences using connectives.**
- First-order logic statements can be divided into two parts:
- Subject: **Subject is the main part of the statement.**
- Predicate: **A predicate can be defined as a relation, which binds two atoms together in a statement.**
- Consider the statement: "x is an integer.", **it consists of two parts, the first part x is the subject of the statement and second part "is an integer," is known as a predicate.**

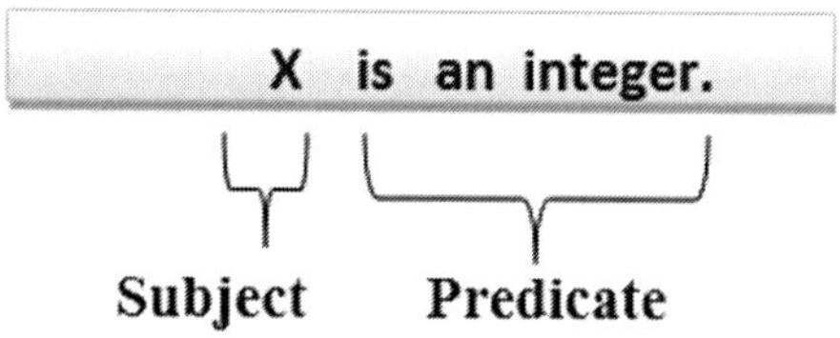

Quantifiers in First-order logic

- A quantifier is a language element which generates quantification, and quantification specifies the quantity of specimen in the universe of discourse.
- These are the symbols that permit to determine or identify the range and scope of the variable in the logical expression.

There are two types of quantifier:

1. Universal Quantifier, (for all, everyone, everything)
2. Existential quantifier, (for some, at least one).

Universal Quantifier

Universal quantifier is a symbol of logical representation, which specifies that the statement within its range is true for everything or every instance of a particular thing.
The Universal quantifier is represented by a symbol ∀,

which resembles an inverted A
If x is a variable, then ∀x is read as:

- **For all x**
- **For each x**
- **For every x.**

Example
All man drink coffee.
Let a variable x which refers to a cat so all x can be represented in UOD as below:
It will be read as: There are all x where x is a man who drink coffee.

Existential Quantifier

Existential quantifiers are the type of quantifiers, which express that the statement within its scope is true for at least one instance of something.

It is denoted by the logical operator ∃, which resembles as inverted E. When it is used with a predicate variable then it is called as an existential quantifier.

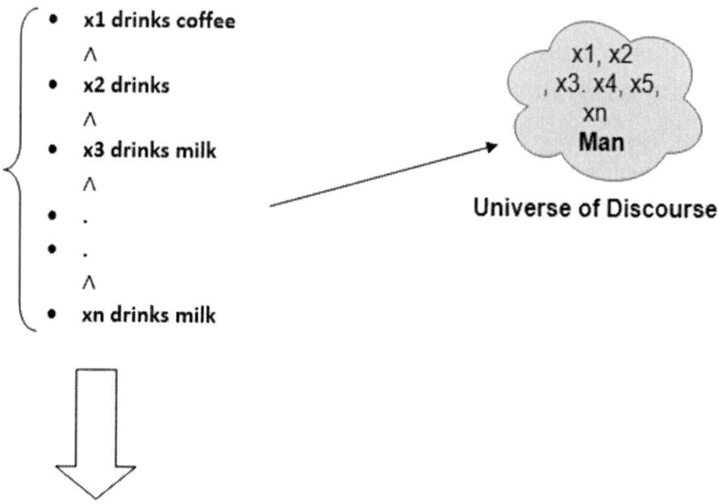

So in shorthand notation, we can write it as :

If x is a variable, then existential quantifier will be ∃x or ∃(x). And it will be read as:

- There exists a 'x.'
- For some 'x.'
- For at least one 'x.'

1.2 Prolog Programming

Prolog stands for **Programming in logic.** It is used in artificial intelligence programming.

Prolog is a **declarative** programming language.

For example: While implementing the solution for a given problem, instead of specifying the ways to achieve a certain goal in a specific situation, user needs to specify about the situation (rules and facts) and the goal (query). After these stages, Prolog interpreter derives the solution.

Prolog is useful in AI, NLP, and databases but useless in other areas such as graphics or numerical algorithms.

Prolog facts

A fact is something that seems to be true. For example: It's raining.

In Prolog, facts are used to form the statements. Facts consist of a specific item or relation between two or more items.

How to convert English to prolog facts using facts and rules?

It is very simple to convert English sentence into Prolog facts. Some examples are explained in the following table.

English Statements	Prolog Facts
Dog is barking	barking(dog)
Jaya likes food if it is delicious.	likes(Jaya, Food):- delicious(Food)

In the above table, the statement 'Dog is barking' is a fact, while the statement 'Jaya likes food if it is delicious' is called rule. In this statement, variable like 'Food' has a first letter in capital, because its value came from previous fact. The symbol ':-' is used to denote that ―Jaya likes delicious food‖.

For example:

?- X is 2 + 1. X = 3 ?

yes

In the above example, **'is'** is used as a special predefined operator.

The basic arithmetic operators are given in the following table

S. No.	Operator	Explanation
1	X+Y	The sum of 'X' and 'Y'
2	X-Y	the difference of 'X' and 'Y'
3	X*Y	The product of 'X' and 'Y'
4	X/Y	The quotient of 'X' and 'Y'
5	X^Y	'X' to the power of 'Y'
6	-X	Negation of 'X'
7	abs(X)	Absolute value of 'X'
8	sqrt(X)	The square root of X
9	sin(X)	The sine of X
10	cos(X)	The cos of X

Operator precedence

- If there is more than one operator in the arithmetic expression such as A-B*C+D, then the prolog decides an order in which the operator should be applied.
- Prolog gives numerical value for each operator, operators with high precedence like '*' and '/' are applied before operators with relatively low precedence values like '+' and '-'.
- Operator with same precedence value ('*' or '/') and ('+' or '-') should be applied from left to right.
- So, the expression A-B*C+D can be written as **A- (B*C)+D**

Matching and Unification in Prolog

Definition: The two terms are said to be matched, if they are equal or if they consist of variables representing the resulting equal terms.

Prolog matches expressions in structural way. So,

?- 3 + 2= 5 no

Note: In prolog '=' means matches with.

Consider the following example,?- X + 3 = 2 * Y

But the following expressions will match because they have same structure.

Expression 1:

?- X + Y = 2 + 3

X = 2

Y = 3

Expression 2:

?- 2 + Y = X + 3

X = 2

Y = 3

Prolog Lists

- Lists are the finite sequence of elements.
- Prolog uses [...] to build a list.
- The notation [X|Y] represents that the first element is X and second element is Y (X is head and Y is tail).

1. Prolog has some special notation for lists:
2. [a] [honda, maruti, renault]
3. [a,b,c) [pen, pencil, notebook]
4. [] represents the empty list.

Example 1: Pattern Matching in Lists

?- [a,b] = [a,X] X = b

but:

?- [a,b] = [X]

no

Example 2

Consider the following lists:

[a, b, c, d, e, f, g]

[apple, pear, bananas, breadfruit] [ ] this is an empty list

Now, consider some comparisons of lists

[a,b,c] matches with [Head|Tail] resulting in Head=a and Tail=[b,c]

[a] matches with [H|T] resulting in H=a and T=[] [a,b,c] matches with [a|T] resulting in T=[b,c] [a,b,c] doesn't match with [b|T]

[] doesn't match with [H|T]

[] match with []. Hence, two empty lists get matched with each other.

Backtracking in Prolog

What is backtracking?

If a person reaches a point where a goal cannot be matched, so he can come back (backtrack) to the last spot, where the choice of matching a specific fact or rule was formed. If this process fails, a person again goes to the nearest previous place where a choice was made. So, this procedure is followed until the goal is achieved.

Example:

Consider the following facts, bird(type (sparrow) name (steve))) bird(type (penguin) name (sweety))) bird(type (penguin)name (jones)))

consider the following query:

?- bird(type (penguin)name(X)

So, prolog will try to match the first query, but this query will not match because sparrow doesn't match with penguin. Then, it will try to find next query to match the fact and succeed with X = sweety. Later, if the query or subgoals are failed, it will go to the saved option and look for more solutions. For example: X = jones

Genetic Learning

Genetic algorithms are the parts of evolutionary computing and are implemented as a computer simulation.

What is biological evolution?

Every organism has a set of rules which describes, how that organism is built, and encoded in the genes of an organism.

- The chromosomes contain hundreds to thousands of genes.
- Features of the organisms depend on the genes and they have several settings. For example: hair color gene may be brown or black.
- The genes and their settings are referred to as a genotype of an organism.
- When two organisms mate, they share their genes, so that the offspring gets half genes from one parent and half genes from the other parent. Hence, this process is called crossover.
- It is possible that gene is mutated in the organism as a completely new feature.
- The genetic algorithm follows the process of nature to solve the problems such as selection, crossover, mutation, and acceptance for evolution.

Following are the steps of genetic algorithm

Step 1: Generate random population of 'n' chromosomes. Step 2: Evaluate the fitness f(x) of each chromosome 'x' present in the population.

Step 3: Follow these steps to create a new population: a. Selection: From population, select two parents fitness according to their fitness.

a. Crossover: Perform crossover using properties of

crossover to form a new offspring.

b. Mutation: Mutate the offspring at each position in chromosome.

c. Accept: Place a new offspring in the new population.

Step 4: Replace: Use new generated population to run an algorithm.

Step 5: Test and Stop, if end condition is satisfied and return the best solution in the current population.

Step 6: Go to step 2.

1.3 Unification

In propsoitional logic it is easy to determine that two literals can not both be true at the same time. Simply look for L and ~L. In predicate logic, this matching process is more complicated, since bindings of variables must be considered.

For example man (john) and man (john) is a contradiction while man (john) and man(Himalayas) is not. Thus in order to determine contradictions we need a matching procedure that compares two literals and discovers whether there exist a set of substitutions that makes them identical. There is a recursive procedure that does this matching. It is called Unification algorithm.

In Unification algorithm each literal is represented as a list, where first element is the name of a predicate and the remaining elements are arguments. The argument may be a single element (atom) or may be another list. For example we can have literals as

(tryassassinate Marcus Caesar)

(tryassassinate Marcus (ruler of Rome))

To unify two literals, first check if their first elements re same. If so proceed. Otherwise they can not be unified. For example the literals (try assassinate Marcus Caesar) (hate Marcus Caesar)

Cannot be Unfied. The unification algorithm recursively matches pairs of elements, one pair at a time. The matching rules are:

a. Different constants, functions or predicates cannot match, whereas identical ones can.
b. A variable can match another variable, any constant or a function or predicate expression, subject to the condition that the function or [predicate expression must not contain any instance of the variable being matched (otherwise it will lead to infinite recursion).
c. The substitution must be consistent. Substituting y for x now and then z for x later is inconsistent. (a substitution y for x written as y/x)

The Unification algorithm is listed below as a procedure UNIFY (L1, L2). It returns a list representing the composition of the substitutions that were performed during the match. An empty list NIL indicates that a match was found without any substitutions. If the list contains a single value F, it indicates that the unification procedure failed.

UNIFY (L1, L2)

1. if L1 or L2 is an atom part of same thing do a. if L1 or L2 are identical then return NIL

b. else if L1 is a variable then do

- if L1 occurs in L2 then return F else return (L2/L1)

c. else if L2 is a variable then do

- if L2 occurs in L1 then return F else return (L1/L2)

else return F.

2. If length (L!) is not equal to length (L2) then return F.

3. Set SUBST to NIL

(at the end of this procedure , SUBST will contain all the substitutions used to unify L1 and L2).

4. For I = 1 to number of elements in L1 do

a. call UNIFY with the i th element of L1 and I'th element of L2, putting the result in S
b. if S = F then return F
c. if S is not equal to NIL then do
(A) apply S to the remainder of both L1 and L2 (B) SUBST := APPEND (S, SUBST) return SUBST.

1.4 Forward Chaining

In artificial intelligence, forward and backward chaining is one of the important topics, but before understanding forward and backward chaining lets first understand that from where these two terms came.

Inference engine

The inference engine is the component of the intelligent system in artificial intelligence, which applies logical rules to the knowledge base to infer new information from known facts. The first inference engine was part of the expert system. Inference engine commonly proceeds in two modes, which are:

- Forward chaining
- Backward chaining

Horn Clause and Definite clause

Horn clause and definite clause are the forms of sentences, which enables knowledge base to use a more restricted and efficient inference algorithm. Logical inference algorithms use forward and backward chaining approaches, which require KB in the form of the **first-order definite clause**.

Definite clause: A clause which is a disjunction of literals with **exactly one positive literal** is known as a definite clause or strict horn clause.

Horn clause: A clause which is a disjunction of literals with **at most one positive literal** is known as horn clause. Hence all the definite clauses are horn clauses.

Example: (¬ p V ¬ q V k). It has only one positive literal k. It is equivalent to $p \wedge q \rightarrow k$.

A. Forward Chaining

Forward chaining is also known as a forward deduction or forward reasoning method when using an inference engine. Forward chaining is a form of reasoning which start with atomic sentences in the knowledge base and applies inference rules (Modus Ponens) in the forward direction to extract more data until a goal is reached.

The Forward-chaining algorithm starts from known facts, triggers all rules whose premises are satisfied, and add their conclusion to the known facts. This process repeats until the problem is solved.

Properties of Forward-Chaining

- It is a down-up approach, as it moves from bottom to top.
- It is a process of making a conclusion based on known facts or data, by starting from the initial state and reaches the goal state.
- Forward-chaining approach is also called as data- driven as we reach to the goal using available data.
- Forward -chaining approach is commonly used in the expert system, such as CLIPS, business, and production rule systems.

Consider the following famous example which we will use in both approaches:

Example

"As per the law, it is a crime for an American to sell weapons to hostile nations. Country A, an enemy of America,

has some missiles, and all the missiles were sold to it by Robert, who is an American citizen."
Prove that **"Robert is criminal."**

To solve the above problem, first, we will convert all the above facts into first-order definite clauses, and then we will use a forward-chaining algorithm to reach the goal.

Facts Conversion into FOL

- It is a crime for an American to sell weapons to hostile nations. (Let's say p, q, and r are variables)

American (p) ∧ weapon(q) ∧ sells (p, q, r) ∧ hostile(r)

→ Criminal(p) ...... (1)

Country A has some missiles. **?p Owns(A, p) ∧**

Missile(p). It can be written in two definite clauses by using Existential Instantiation, introducing new

Constant T1.

Owns(A, T1) ...... (2)

Missile(T1) ...... (3)

- All of the missiles were sold to country A by Robert.

?p Missiles(p) ∧ Owns (A, p) → Sells (Robert, p,)

...... (4)

□ Missiles are weapons.

Missile(p) → Weapons (p)

□ Enemy of America is known as hostile.

Enemy(p, America) →Hostile(p) ...... (6)

Country A is an enemy of America.

Enemy (A, America) ...... (7)

Robert is American

American(Robert). ...... (8)

Forward chaining proof

Step-1:

In the first step we will start with the known facts and will choose the sentences which do not have implications, such as: **American(Robert), Enemy(A, America), Owns(A, T1), and Missile(T1)**. All these facts will be represented as below.

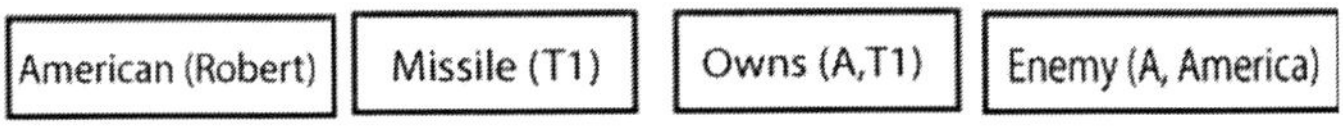

Step-2:

At the second step, we will see those facts which infer from available facts and with satisfied premises.

Rule-(1) does not satisfy premises, so it will not be added in the first iteration.

Rule-(2) and (3) are already added.

Rule-(4) satisfy with the substitution {p/T1}, **so Sells (Robert, T1, A)** is added, which infers from the conjunction of Rule (2) and (3).

Rule-(6) is satisfied with the substitution (p/A), so Hostile(A) is added and which infers from Rule-(7).

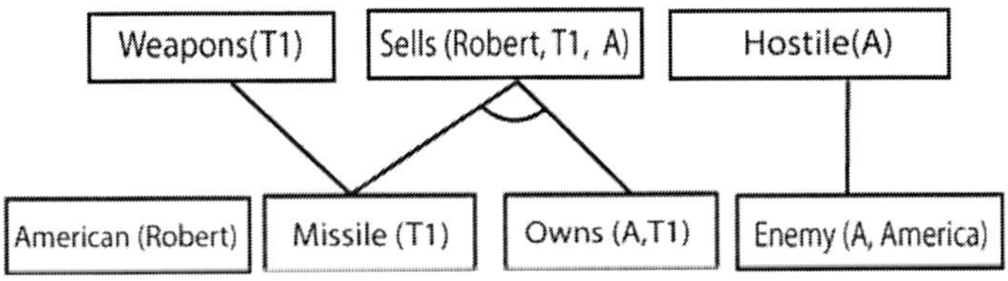

Step-3

At step-3, as we can check Rule-(1) is satisfied with the substitution **{p/Robert, q/T1, r/A}, so we can add Criminal(Robert)** which infers all the available facts. And hence we reached our goal statement.

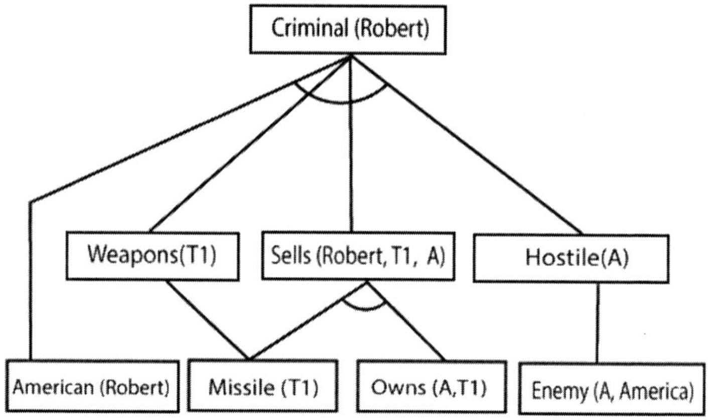

3.5 Backward Chaining

Backward-chaining is also known as a backward deduction or backward reasoning method when using an inference engine. A backward chaining algorithm is a form of reasoning, which starts with the goal and works backward, chaining through rules to find known facts that support the goal.

Properties of backward chaining

- It is known as a top-down approach.
- Backward-chaining is based on modus ponens inference rule.
- In backward chaining, the goal is broken into sub-goal or sub-goals to prove the facts true.It is called a goal-driven approach, as a list of goals decides which rules are selected and used.
- Backward -chaining algorithm is used in game theory, automated theorem proving tools, inference engines, proof assistants, and various AI applications.
- The backward-chaining method mostly used a **depth-first search** strategy for proof.

Example

In backward-chaining, we will use the same above example, and will rewrite all the rules.

□ **American (p) ∧ weapon(q) ∧ sells (p, q, r) ∧ hostile(r)**

→ Criminal(p) ...... (1)

Owns(A, T1) ...... (2)

□ **Missile(T1) ...... (3)**

□ **?p Missiles(p) ∧ Owns (A, p) → Sells (Robert, p, A)**

...... (4)

- □ **Missile(p) → Weapons (p)** ...... **(5)**
- □ **Enemy(p, America) →Hostile(p)** ...... **(6)**
- □ **Enemy (A, America)** ...... **(7)**
- □ **American(Robert).** ...... **(8)**

Backward-Chaining proof

In Backward chaining, we will start with our goal predicate, which is **Criminal(Robert)**, and then infer further rules.

Step-1:

At the first step, we will take the goal fact. And from the goal fact, we will infer other facts, and at last, we will prove those facts true. So our goal fact is "Robert is Criminal," so following is the predicate of it.

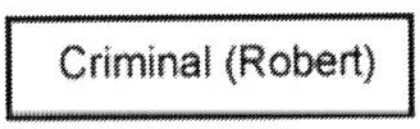

Step-2

At the second step, we will infer other facts form goal fact which satisfies the rules. So as we can see in Rule-1, the goal predicate Criminal (Robert) is present with substitution

{Robert/P}. So we will add all the conjunctive facts below the first level and will replace p with Robert.

Here we can see American (Robert) is a fact, so it is proved here.

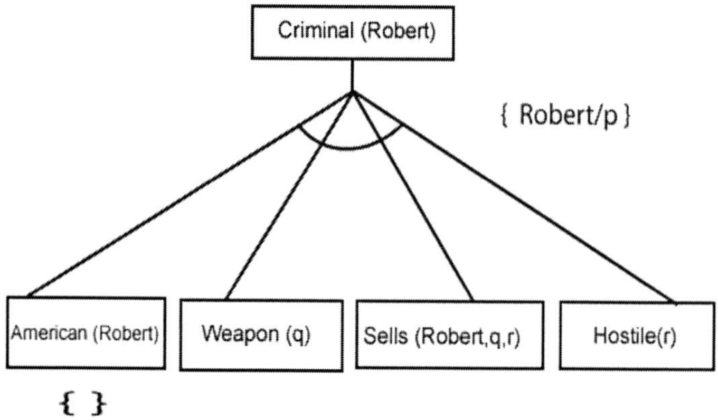

Step-3

t At step-3, we will extract further fact Missile(q) which infer from Weapon(q), as it satisfies Rule-(5). Weapon (q) is also true with the substitution of a constant T1 at q.

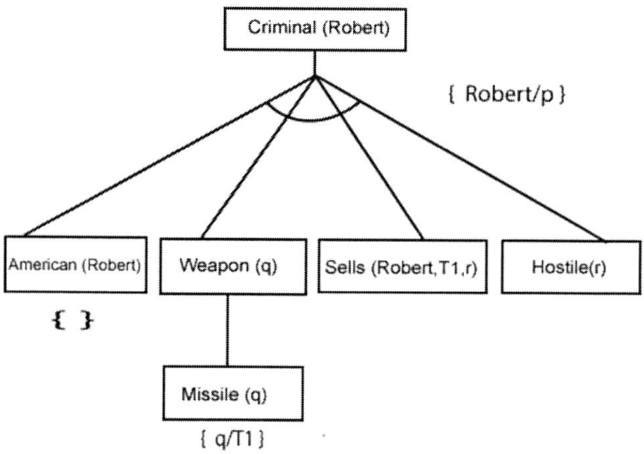

Step-4:

At step-4, we can infer facts Missile(T1) and Owns(A, T1) form Sells(Robert, T1, r) which satisfies the **Rule- 4**, with the substitution of A in place of r. So these two statements are proved here.

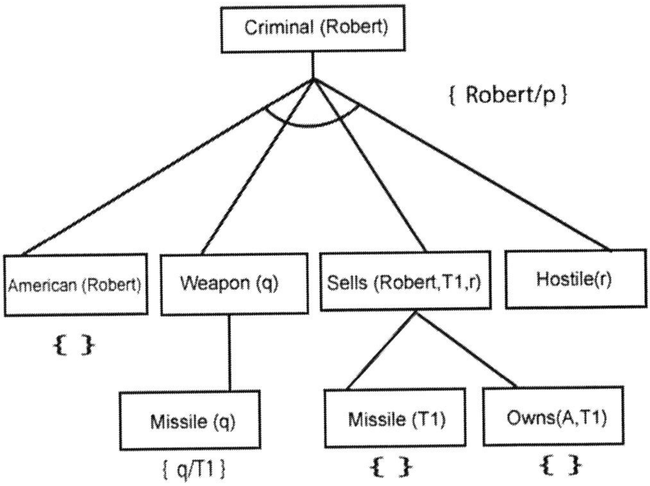

Step-5

At step-5, we can infer the fact **Enemy(A, America)** from **Hostile(A)** which satisfies Rule- 6. And hence all the statements are proved true using backward chaining.

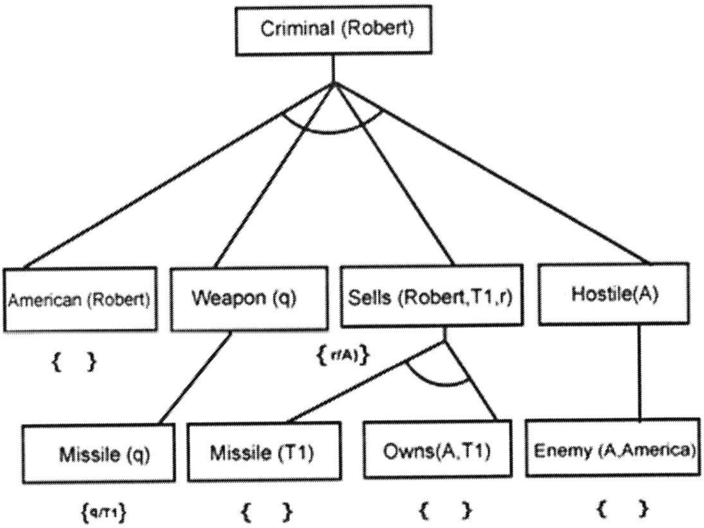

Difference between backward chaining and forward chaining

Following is the difference between the forward chaining and backward chaining

- Forward chaining as the name suggests, start from the known facts and move forward by applying inference rules to extract more data, and it continues until it reaches to the goal, whereas backward chaining starts from the goal, move backward by using inference rules to determine the facts that satisfy the goal.
- Forward chaining is called a **data-driven** inference technique, whereas backward chaining is called a **goal-driven** inference technique.
- Forward chaining is known as the **down-up** approach, whereas backward chaining is known as a **top- down** approach.
- Forward chaining uses **breadth-first search** strategy, whereas backward chaining uses **depth-first search** strategy.
- Forward and backward chaining both applies **Modus ponens** inference rule.

S. No.	Forward Chaining	Backward Chaining
1.	Forward chaining starts from known facts and applies inference rule to extract more data unit it reaches to the goal.	Backward chaining starts from the goal and works backward through inference rules to find the required facts that support the goal.
2.	It is a bottom-up approach	It is a top-down approach

3.	Forward chaining is known as data-driven inference technique as we reach to the goal using the available data.	Backward chaining is known as goal-driven technique as we start from the goal and divide into sub-goal to extract the facts.
4.	Forward chaining reasoning applies a breadth-first search strategy.	Backward chaining reasoning applies a depth-first search strategy.
5.	Forward chaining tests for all the available rules	Backward chaining only tests for few required rules.
6.	Forward chaining is suitable for the planning, monitoring, control, and interpretation application.	Backward chaining is suitable for diagnostic, prescription, and debugging application.
7.	Forward chaining can generate an infinite number of possible conclusions.	Backward chaining generates a finite number of possible conclusions.
8.	It operates in the forward direction.	It operates in the backward direction.
9.	Forward chaining is aimed for any conclusion.	Backward chaining is only aimed for the required data.

- Forward chaining can be used for tasks such as **planning, design process monitoring, diagnosis, and classification**, whereas backward chaining can be used for **classification and diagnosis tasks**.
- Forward chaining can be like an exhaustive search, whereas backward chaining tries to avoid the unnecessary path of reasoning.
- In forward-chaining there can be various ASK questions from the knowledge base, whereas in backward chaining

there can be fewer ASK questions.

- Forward chaining is slow as it checks for all the rules, whereas backward chaining is fast as it checks few required rules only

3.6. Resolution

Resolution is a theorem proving technique that proceeds by building refutation proofs, i.e., proofs by contradictions. It was invented by a Mathematician John Alan Robinson in the year 1965. Resolution is used, if there are various statements are given, and we need to prove a conclusion of those statements. Unification is a key concept in proofs by resolutions. Resolution is a single inference rule which can efficiently operate on the **conjunctive normal form or clausal form.**

Clause: Disjunction of literals (an atomic sentence) is called a **clause.** It is also known as a unit clause.

Conjunctive Normal Form: A sentence represented as a conjunction of clauses is said to be **conjunctive normal form** or **CNF.**

The resolution inference rule

The resolution rule for first-order logic is simply a lifted version of the propositional rule. Resolution can resolve two clauses if they contain complementary literals, which are assumed to be standardized apart so that they share no variables.

$$\frac{l_1 \vee \ldots\ldots \vee l_k, \quad m_1 \vee \ldots\ldots \vee m_n}{SUBST(\theta, l_1 \vee \ldots\ldots \vee l_{i-1} \vee l_{i+1} \vee \ldots \vee l_k \vee m_1 \vee \ldots\ldots \vee m_{j-1} \vee m_{j+1} \vee \ldots\ldots \vee m_n)}$$

Where **li** and **mj** are complementary literals.

This rule is also called the **binary resolution rule** because it only resolves exactly two literals.

Example

We can resolve two clauses which are given below:

[Animal (g(x) V Loves (f(x), x)] and [¬Loves(a, b) V ¬Kills(a, b)]

Where two complimentary literals are: **Loves (f(x), x) and ¬Loves (a, b)**

These literals can be unified with unifier θ= **[a/f(x), and b/x]** , and it will generate a resolvent clause:

[Animal (g(x) V ¬Kills(f(x), x)].

Steps for Resolution

1. Conversion of facts into first-order logic.
2. Convert FOL statements into CNF
3. Negate the statement which needs to prove (proof by contradiction)
4. Draw resolution graph (unification).
5. To better understand all the above steps, we will take an example in which we will apply resolution.

Example

a. John likes all kind of food.
b. Apple and vegetable are food
c. Anything anyone eats and not killed is food.
d. Anil eats peanuts and still alive
e. Harry eats everything that Anil eats.

Prove by resolution that:

a. John likes peanuts.

Step-1: Conversion of Facts into FOL

In the first step we will convert all the given statements into its first order logic.

a. ∀x: food(x) → likes(John, x)

b. food(Apple) Λ food(vegetables)

c. ∀x ∀y: eats(x, y) Λ ¬ killed(x) → food(y)

d. eats (Anil, Peanuts) Λ alive(Anil).

e. ∀x : eats(Anil, x) → eats(Harry, x)

f. ∀x: ¬ killed(x) → alive(x) } added predicates.

g. ∀x: alive(x) →¬ killed(x)

h. likes(John, Peanuts)

Step-2: Conversion of FOL into CNF

In First order logic resolution, it is required to convert the FOL into CNF as CNF form makes easier for resolution proofs.

Eliminate all implication (→) and rewrite

1. ∀x ¬ food(x) V likes(John, x)
2. food(Apple) Λ food(vegetables)
3. ∀x ∀y ¬ [eats(x, y) Λ ¬ killed(x)] V food(y)
4. eats (Anil, Peanuts) Λ alive(Anil)
5. ∀x ¬ eats(Anil, x) V eats(Harry, x)
6. ∀x¬ [¬ killed(x)] V alive(x)
7. ∀x ¬ alive(x) V ¬ killed(x)
8. likes(John, Peanuts).

Move negation (¬)inwards and rewrite

1. ∀x ¬ food(x) V likes(John, x)
2. food(Apple) Λ food(vegetables)
3. ∀x ∀y ¬ eats(x, y) V killed(x) V food(y)
4. eats (Anil, Peanuts) Λ alive(Anil)
5. ∀x ¬ eats(Anil, x) V eats(Harry, x)

6. ∀x ¬killed(x)] V alive(x)
7. ∀x ¬ alive(x) V ¬ killed(x)
8. likes(John, Peanuts).

Rename variables or standardize variables

1. ∀x ¬ food(x) V likes(John, x)
2. food(Apple) Λ food(vegetables)
3. ∀y ∀z ¬ eats(y, z) V killed(y) V food(z)
4. eats (Anil, Peanuts) Λ alive(Anil)
5. ∀w¬ eats(Anil, w) V eats(Harry, w)
6. ∀g ¬killed(g)] V alive(g)
7. ∀k ¬ alive(k) V ¬ killed(k)
8. likes(John, Peanuts).

Eliminate existential instantiation quantifier by elimination.

In this step, we will eliminate existential quantifier ∃, and this process is known as **Skolemization**. But in this example problem since there is no existential quantifier so all the statements will remain same in this step.

Drop Universal quantifiers

In this step we will drop all universal quantifier since all the statements are not implicitly quantified so we don't need
it.

1. ¬ food(x) V likes(John, x)
2. food(Apple)
3. food(vegetables)
4. ¬ eats(y, z) V killed(y) V food(z)
5. eats (Anil, Peanuts)

6. alive(Anil)
7. ¬ eats(Anil, w) V eats(Harry, w)
8. killed(g) V alive(g)
9. ¬ alive(k) V ¬ killed(k)
10. likes(John, Peanuts).

Distribute conjunction ∧ over disjunction ¬.

This step will not make any change in this problem.

Step-3: Negate the statement to be proved

In this statement, we will apply negation to the conclusion statements, which will be written as ¬likes(John, Peanuts)

Step-4: Draw Resolution graph

Now in this step, we will solve the problem by resolution tree using substitution. For the above problem, it will be given as follows:

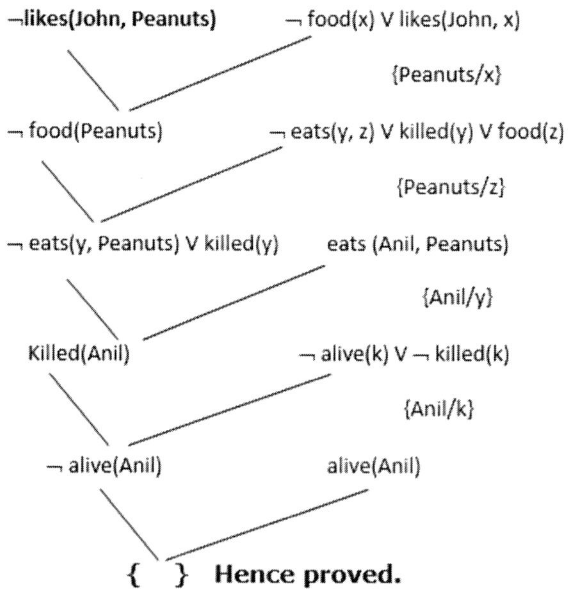

Hence the negation of the conclusion has been proved as a complete contradiction with the given set of statements.

Explanation of Resolution graph

- In the first step of resolution graph, **¬likes(John, Peanuts)** , and **likes(John, x)** get resolved(canceled) by substitution of **{Peanuts/x}**, and we are left with **¬ food(Peanuts)**
- In the second step of the resolution graph, **¬ food(Peanuts)** , and **food(z)** get resolved (canceled) by substitution of **{ Peanuts/z}**, and we are left with **¬ eats(y, Peanuts) V killed(y)** .
- In the third step of the resolution graph, **¬ eats(y, Peanuts)** and **eats (Anil, Peanuts)** get resolved by substitution **{Anil/y}**, and we are left with **Killed(Anil)** .
- In the fourth step of the resolution graph, **Killed(Anil)** and **¬ killed(k)** get resolve by substitution **{Anil/k}**, and we are left with **¬ alive(Anil)** .
- In the last step of the resolution graph **¬alive(Anil)** and **alive(Anil)** get resolved.

3.8 Knowledge Representation

Humans are best at understanding, reasoning, and interpreting knowledge. Human knows things, which is knowledge and as per their knowledge they perform various actions in the real world. **But how machines do all these things comes under knowledge representation and reasoning.** Hence we can describe Knowledge representation as following:

- Knowledge representation and reasoning (KR, KRR) is the part of Artificial intelligence which concerned with AI agents thinking and how thinking contributes to intelligent behavior of agents.
- It is responsible for representing information about the real world so that a computer can understand and can utilize this knowledge to solve the complex real world problems such as diagnosis a medical condition or

communicating with humans in natural language.
- It is also a way which describes how we can represent knowledge in artificial intelligence. Knowledge representation is not just storing data into some database, but it also enables an intelligent machine to learn from that knowledge and experiences so that it can behave intelligently like a human.

What to Represent

Following are the kind of knowledge which needs to be represented in AI systems:

- **Object:** All the facts about objects in our world domain. E.g., Guitars contains strings, trumpets are brass instruments.
- **Events:** Events are the actions which occur in our world.
- **Performance:** It describe behavior which involves knowledge about how to do things.
- **Meta-knowledge:** It is knowledge about what we know.
- **Facts:** Facts are the truths about the real world and what we represent.
- **Knowledge-Base:** The central component of the knowledge-based agents is the knowledge base. It is represented as KB. The Knowledgebase is a group of the Sentences (Here, sentences are used as a technical term and not identical with the English language).

Knowledge: Knowledge is awareness or familiarity gained by experiences of facts, data, and situations. Following are the types of knowledge in artificial intelligence:

Types of knowledge

Following are the various types of knowledge

1. Declarative Knowledge

- Declarative knowledge is to know about something.
- It includes concepts, facts, and objects.
- It is also called descriptive knowledge and expressed in declarative sentences.
- It is simpler than procedural language.

2. Procedural Knowledge

- It is also known as imperative knowledge.
- Procedural knowledge is a type of knowledge which is responsible for knowing how to do something.
- It can be directly applied to any task.
- It includes rules, strategies, procedures, agendas, etc.
- Procedural knowledge depends on the task on which it can be applied.

3. Meta-knowledge

- Knowledge about the other types of knowledge is called Meta-knowledge.

4. Heuristic knowledge:

- Heuristic knowledge is representing knowledge of some experts in a filed or subject.
- Heuristic knowledge is rules of thumb based on previous experiences, awareness of approaches, and which are good to work but not guaranteed.

5. Structural knowledge

- Structural knowledge is basic knowledge to problem-solving.
- It describes relationships between various concepts such as kind of, part of, and grouping of something.
- It describes the relationship that exists between concepts or objects.

The relation between knowledge and intelligence

Knowledge of real-worlds plays a vital role in intelligence and same for creating artificial intelligence. Knowledge plays an important role in demonstrating intelligent behavior in AI agents. An agent is only able to accurately act on some input when he has some knowledge or experience about that input.

Let's suppose if you met some person who is speaking in a language which you don't know, then how you will able to act on that. The same thing applies to the intelligent behavior of the agents.

As we can see in below diagram, there is one decision maker which act by sensing the environment and using knowledge. But if the knowledge part will not present then, it cannot display intelligent behavior.

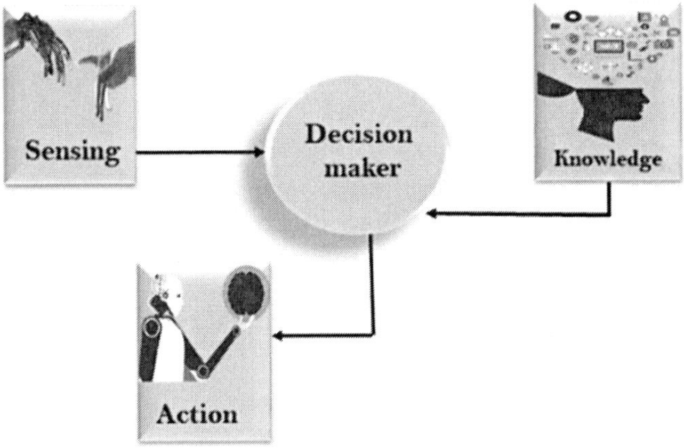

An Artificial intelligence system has the following components for displaying intelligent behavior:

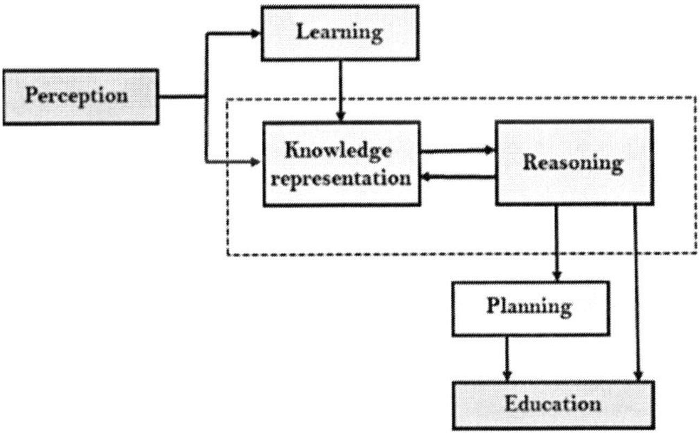

- Perception
- Learning
- Knowledge Representation and Reasoning
- Planning
- Execution

The above diagram is showing how an AI system can interact with the real world and what components help it to show intelligence. AI system has Perception component by which it retrieves information from its environment. It can be visual, audio or another form of sensory input. The learning component is responsible for learning from data captured by Perception comportment. In the complete cycle, the main components are knowledge representation and Reasoning. These two components are involved in showing the intelligence in machine-like humans. These two components are independent with each other but also coupled together. The planning and execution depend on analysis of Knowledge representation and reasoning.

Approaches to knowledge representation

There are mainly four approaches to knowledge representation, which are given below:

1. Simple relational knowledge

- It is the simplest way of storing facts which uses the relational method, and each fact about a set of the object is set out systematically in columns.
- This approach of knowledge representation is famous
- in database systems where the relationship between different entities is represented.
- This approach has little opportunity for inference.

Example: The following is the simple relational knowledge representation.

Player	**Weight**	**Age**
Player1	65	23
Player2	58	18
Player3	75	24

2. Inheritable knowledge:

- In the inheritable knowledge approach, all data must be stored into a hierarchy of classes.
- All classes should be arranged in a generalized form or a hierarchal manner.

- In this approach, we apply inheritance property.
- Elements inherit values from other members of a class.
- This approach contains inheritable knowledge which shows a relation between instance and class, and it is called instance relation.
- Every individual frame can represent the collection of attributes and its value.
- In this approach, objects and values are represented in Boxed nodes.
- We use Arrows which point from objects to their values.

Example

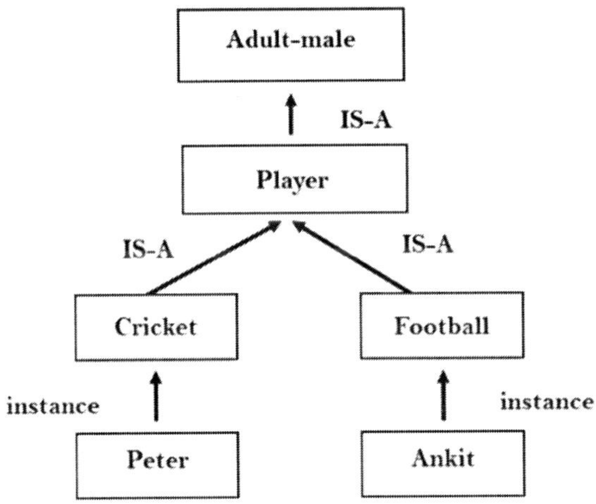

3. Inferential knowledge

- Inferential knowledge approach represents knowledge in the form of formal logics.
- This approach can be used to derive more facts.
- It guaranteed correctness.

Example: Let's suppose there are two statements:

1. Marcus is a man
2. All men are mortal

3. Then it can represent as;

man(Marcus)

∀x = man (x) ----------> mortal (x)s

4. Procedural knowledge:

- Procedural knowledge approach uses small programs and codes which describes how to do specific things, and how to proceed.
- In this approach, one important rule is used which is **If-Then rule**.
- In this knowledge, we can use various coding languages such as **LISP language** and **Prolog language**.
- We can easily represent heuristic or domain- specific knowledge using this approach.
- But it is not necessary that we can represent all cases in this approach.

Requirements for knowledge Representation system

A good knowledge representation system must possess the following properties.

1. Representational Accuracy

KR system should have the ability to represent all kind of required knowledge.

2. Inferential Adequacy

KR system should have ability to manipulate the representational structures to produce new knowledge corresponding to existing structure.

3. Inferential Efficiency

The ability to direct the inferential knowledge mechanism into the most productive directions by storing appropriate guides.

4. Acquisitional efficiency

The ability to acquire the new knowledge easily using automatic methods.

Techniques of knowledge representation

There are mainly four ways of knowledge representation which are given as follows:

1. Logical Representation
2. Semantic Network Representation
3. Frame Representation
4. Production Rules

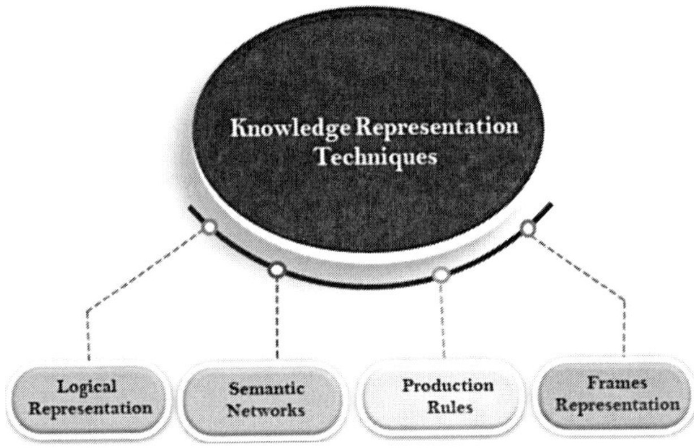

1. Logical Representation

Logical representation is a language with some concrete rules which deals with propositions and has no ambiguity in representation. Logical representation means drawing a conclusion based on various conditions. This representation lays down some important communication rules. It consists of precisely defined syntax and semantics which supports the sound inference. Each sentence can be translated into logics using syntax and semantics.

Syntax

- Syntaxes are the rules which decide how we can construct legal sentences in the logic.
- It determines which symbol we can use in knowledge representation.
- How to write those symbols.

Semantics

- Semantics are the rules by which we can interpret the sentence in the logic.
- Semantic also involves assigning a meaning to each sentence.

Logical representation can be categorised into mainly two logics

- Propositional Logics
- Predicate logics

Advantages of logical representation

1. Logical representation enables us to do logical reasoning.
2. Logical representation is the basis for the programming languages.

Disadvantages of logical Representation

1. Logical representations have some restrictions and are challenging to work with.
2. Logical representation technique may not be very natural, and inference may not be so efficient.

2. Semantic Network Representation

Semantic networks are alternative of predicate logic for knowledge representation. In Semantic networks, we can represent our knowledge in the form of graphical networks. This network consists of nodes representing objects and arcs which describe the relationship between those objects. Semantic networks can categorize the object in different forms and can also link those objects. Semantic networks are easy to understand and can be easily extended.

This representation consist of mainly two types of relations:

- IS-A relation (Inheritance)
- Kind-of-relation

Example: Following are some statements which we need to represent in the form of nodes and arcs.

Statements

1. Jerry is a cat.
2. Jerry is a mammal
3. Jerry is owned by Priya.
4. Jerry is brown colored.
5. All Mammals are animal.

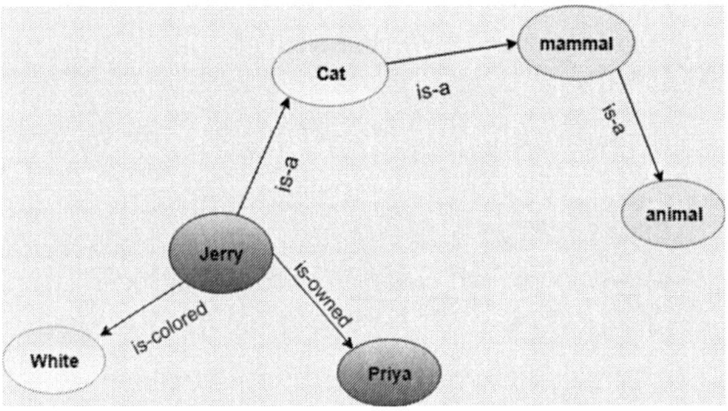

In the above diagram, we have represented the different type of knowledge in the form of nodes and arcs. Each object is connected with another object by some relation.

Drawbacks in Semantic representation

1. Semantic networks take more computational time at runtime as we need to traverse the complete network tree to answer some questions. It might be possible in the worst case scenario that after traversing the entire tree, we find that the solution does not exist in this network.
2. Semantic networks try to model human-like memory (Which has 1015 neurons and links) to store the information, but in practice, it is not possible to build such a vast semantic network.

3. These types of representations are inadequate as they do not have any equivalent quantifier, e.g., for all, for some, none, etc.
4. Semantic networks do not have any standard definition for the link names.
5. These networks are not intelligent and depend on the creator of the system.

Advantages of Semantic network

1. Semantic networks are a natural representation of knowledge.
2. Semantic networks convey meaning in a transparent manner.
3. These networks are simple and easily understandable.

3. Frame Representation

A frame is a record like structure which consists of a collection of attributes and its values to describe an entity in the world. Frames are the AI data structure which divides knowledge into substructures by representing stereotypes situations. It consists of a collection of slots and slot values. These slots may be of any type and sizes. Slots have names and values which are called facets.

Facets: The various aspects of a slot is known as **Facets**. Facets are features of frames which enable us to put constraints on the frames. Example: IF-NEEDED facts are called when data of any particular slot is needed. A frame may consist of any number of slots, and a slot may include any number of facets and facets may have any number of values. A frame is also known as **slot-filter knowledge representation** in artificial intelligence.

Frames are derived from semantic networks and later evolved into our modern-day classes and objects. A single frame is not much useful. Frames system consist of a collection of frames which are connected. In the frame, knowledge about an object or event can be stored together in the knowledge base. The frame is a type of technology which is widely used in various applications including Natural language processing and machine visions.

Example: 1

Let's take an example of a frame for a book

Slots	Filters
Title	Artificial Intelligence
Genre	Computer Science
Author	Peter Norvig
Edition	Third Edition
Year	1996
Page	1152

Example 2

Let's suppose we are taking an entity, Peter. Peter is an engineer as a profession, and his age is 25, he lives in city London, and the country is England. So following is the frame representation for this:

Slots	Filter
Name	Peter
Profession	Doctor
Age	25
Marital status	Single
Weight	78

Advantages of frame representation

1. The frame knowledge representation makes the programming easier by grouping the related data.

3. The frame representation is comparably flexible and used by many applications in AI.
4. It is very easy to add slots for new attribute and relations.
5. It is easy to include default data and to search for missing values.
6. Frame representation is easy to understand and visualize.

Disadvantages of frame representation

1. In frame system inference mechanism is not be easily processed.
2. Inference mechanism cannot be smoothly proceeded by frame representation.
3. Frame representation has a much generalized approach.

4. Production Rules

Production rules system consist of (**condition, action**) pairs which mean, "If condition then action". It has mainly three parts:

- The set of production rules
- Working Memory
- The recognize-act-cycle

In production rules agent checks for the condition and if the condition exists then production rule fires and corresponding action is carried out. The condition part of the rule determines which rule may be applied to a problem. And the action part carries out the associated problem-solving steps. This complete process is called a recognize-act cycle. The working memory contains the description of the current state of problems-solving and rule can write knowledge to the working memory. This knowledge match and may fire other rules.

If there is a new situation (state) generates, then multiple production rules will be fired together, this is called conflict set. In this situation, the agent needs to select a rule from these sets, and it is called a conflict resolution.

Example

- IF (at bus stop AND bus arrives) THEN action (get into the bus)
- IF (on the bus AND paid AND empty seat) THEN action (sit down).
- IF (on bus AND unpaid) THEN action (pay charges).
- IF (bus arrives at destination) THEN action (get down from the bus).

Advantages of Production rule

1. The production rules are expressed in natural language.
2. The production rules are highly modular, so we can easily remove, add or modify an individual rule.

Disadvantages of Production rule

1. Production rule system does not exhibit any learning capabilities, as it does not store the result of the problem for the future uses.
2. During the execution of the program, many rules may be active hence rule-based production systems are inefficient.

3.8 Ontological Engineering

Building large knowledge-based systems is complex because

- Knowledge often comes from multiple sources and must be integrated. Moreover, these sources may not have the same division of the world. Often knowledge comes from different fields that have their own distinctive terminology and divide the world according to their own needs.

- Systems evolve over time and it is difficult to anticipate all future distinctions that should be made.
- The people involved in designing a knowledge base must choose what individuals and relationships to represent. The world is not divided into individuals; that is something done by intelligent agents to understand the world. Different people involved in a knowledge-based system should agree on this division of the world.
- It is difficult to remember what your own notation means, let alone to discover what someone else's notation means. This has two aspects: given a symbol used in the computer, determining what it means; given a concept in someone's mind, determining what symbol they should use; that is, determining whether the concept has been used before and, if it has, discovering what notation has been used for it.

Ontologies and Knowledge Sharing

Building large knowledge-based systems is complex because

- Knowledge often comes from multiple sources and must be integrated. Moreover, these sources may not have the same division of the world. Often knowledge comes from different fields that have their own distinctive terminology and divide the world according to their own needs.
- Systems evolve over time and it is difficult to anticipate all future distinctions that should be made.
- The people involved in designing a knowledge base must choose what individuals and relationships to represent. The world is not divided into individuals; that is something done by intelligent agents to understand the world. Different people involved in a knowledge-based system should agree on this division of the world.

- It is difficult to remember what your own notation means, let alone to discover what someone else's notation means. This has two aspects:
- given a symbol used in the computer, determining what it means;
- given a concept in someone's mind, determining what symbol they should use; that is, determining whether the concept has been used before and, if it has, discovering what notation has been used for it.

The Semantic Web

The **semantic web** is a way to allow machine-interpretable knowledge to be distributed on the World Wide Web. Instead of just serving HTML pages that are meant to be read by humans, web sites will also provide information that can be used by computers.

At the most basic level, **XML** (the Extensible Markup Language) provides a syntax designed to be machine readable, but which is possible for humans to read. It is a text-based language, where items are tagged in a hierarchical manner. The syntax for XML can be quite complicated, but at the simplest level, the scope of a tag is either in the form ⟨*tag*.../⟩, or in the form ⟨*tag*...⟩...⟨/*tag*⟩.

A **URI** (a **Uniform Resource Identifier**) is used to uniquely identify a resource. A **resource** is anything that can be uniquely identified. A URI is a string that refers to a resource, such as a web page, a person, or a corporation. Often URIs use the syntax of web addresses.

RDF (the **Resource Description Framework**) is a language built on XML, providing individual-property-value triples.

RDF-S (RDF Schema) lets you define resources (and so also properties) in terms of other resources (e.g., using *subClassOf*). RDF-S also lets you restrict the domain and range of properties and provides containers (sets, sequences, and alternatives - one of which must be true).

RDF allows sentences in its own language to be reified.

This means that it can represent arbitrary logical formulas and so is not decidable in general. Undecidability is not necessarily a bad thing; it just means that you cannot put a bound on the time a computation may take. Simple logic programs with function symbols and virtually all programming languages are undecidable.

OWL (the **Web Ontology Language**) is an ontology language for the World Wide Web. It defines some classes and properties with a fixed interpretation that can be used for describing classes, properties, and individuals. It has built-in mechanisms for equality of individuals, classes, and properties, in addition to restricting domains and ranges of properties and other restrictions on properties (e.g., transitivity, cardinality).

There have been some efforts to build large universal ontologies, such as **cyc** (www.cyc.com), but the idea of the semantic web is to allow communities to converge on ontologies. Anyone can build an ontology. People who want to develop a knowledge base can use an existing ontology or develop their own ontology, usually built on existing ontologies. Because it is in their interest to have semantic interoperability, companies and individuals should tend to converge on standard ontologies for their domain or to develop mappings from their ontologies to others' ontologies.

To share and communicate knowledge, it is important to be able to come up with a common vocabulary and an agreed- on meaning for that vocabulary.

A **conceptualization** is a mapping between symbols

used in the computer (i.e., the vocabulary) and the individuals and relations in the world. It provides a particular abstraction of the world and notation for that abstraction. A conceptualization for small knowledge bases can be in the head of the designer or specified in natural language in the documentation. This informal specification of a conceptualization does not scale to larger systems where the conceptualization must be shared.

In philosophy, **ontology** is the study of what exists. In AI, an **ontology** is a specification of the meanings of the symbols in an information system. That is, it is a specification of a conceptualization. It is a specification of what individuals and relationships are assumed to exist and what terminology is used for them. Typically, it specifies what types of individuals will be modeled, specifies what properties will be used, and gives some axioms that restrict the use of that vocabulary.

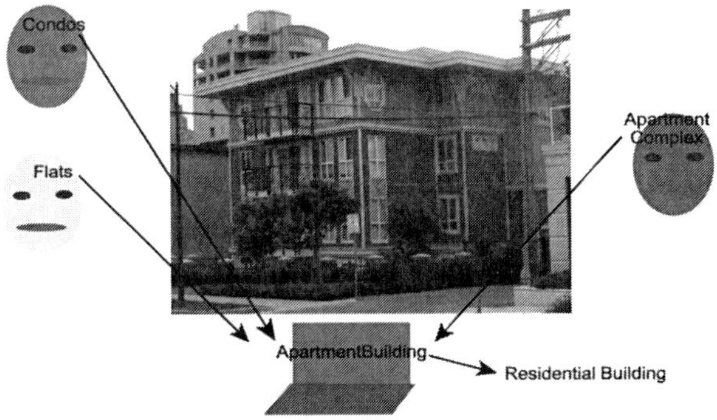

An ontology of individuals that could appear on a map could specify that the symbol "Apartment Building" will represent apartment buildings. The ontology will not *define* an apartment building, but it will describe it well enough so that others can understand the definition. We want other people, who may be inclined to use different symbols, to be able to use the ontology to find the appropriate symbol to use. Multiple people are able to use the symbol consistently. An ontology should also enable a person to verify what a symbol means. That is, given a concept, they want to be able to find the symbol, and, given the symbol, they want to be able to determine what it means.

An ontology may give axioms to restrict the use of some symbol. For example, it may specify that apartment buildings are buildings, which are human-constructed artifacts. It may give some restriction on the size of buildings so that shoeboxes cannot be buildings or that cities cannot be buildings. It may state that a building cannot be at two geographically dispersed locations at the same time (so if you take off some part of the building and move it to a different location, it is no longer a single building). Because apartment buildings are buildings, these restrictions also apply to apartment buildings.

Ontologies are usually written independently of a particular application and often involve a community to agree on the meanings of symbols. An ontology consists of

- a vocabulary of the categories of the things (both classes and properties) that a knowledge base may want to represent;
- an organization of the categories, for example into an inheritance hierarchy using *subClasOf* or *subPropertyOf*, or using ; and
- a set of axioms restricting the meanings of some of the symbols to better reflect their meaning - for example, that some property is transitive, or that the domain and range are restricted, or that there are some restriction on the number of values a property can take for each individual. Sometimes relationships are defined in terms of more primitive relationships but, ultimately, the relationships are grounded out into primitive relationships that are not actually defined.

3.9 Categories and Objects

- Organizing the objects in categories.
- Some reasoning takes places at the level of categories.
- "I want to eat an apple."
- Apple(x), or Member(x, Apple) and Subset(Apple, Food).
- The categories form a hierarchy where each class inherits the properties of the parent, or simply a network (apples are both food and fruit).
- The categories of a class compose a taxonomy.

Partitioning

- Two or more categories are disjoint if they are mutually exclusive (male/female).
- A decomposition of a class into categories is called exhaustive if each object of the class must belong to at least one category
- living = {animal, vegetable, fungi, bacteria}
- A partition is a an exhaustive decomposition of a class into disjoint subsets.
- student = {undergraduate, graduate}
- Related problem: clustering. Given a set of objects, find an appropriate category for each of them. The categories may not be predefined.

Attributes

- Composition: PartOf(nose, face).
- Some categories can be defined as rules:
- Biped(x) => ∃b1, b2 (Leg(b1) & Leg(b2) & PartOf(b1, x) & PartOf(b2, x) & !equal(b1, b2).
- Prime(n) <=> (Number(n) & n != 1 & ∀d (equal(n%d, 0) => (equal(d, 1) | (equal(d, n))))
- Measurements: Length(x), Price(x), etc. These are functions, not predicates.

- Things: 3 apples. Stuff: water, more water, some quantity of water.
- Situations
- Situation: a predicate that is true at a particular moment in time. They may also have a specific location. A particular state of the system.
- Drives(Mary, car, I90, 03/15/096)
- Fluent: predicates and functions that can change from a moment to another. Happy(x).
- Atemporal or eternal predicates: Warrior(A) will be true for as long as the object A exists. Alive(A) may change from a moment to the next.
- Actions
- Predicates that represent changes in the environment.
- The involve at least a subject, often one or more objects.
- eat(x, y), sleep(x), talk(x), say(x, y)
- When an action is part of a situation, it can result in other predicates becoming true at a later moment in time.
- (Eat(x, y, t) & Apple(y)) => !Hungry(x, t+1)
- Projection: the agent can use some rules to predict the outcome of its possible actions and choose the appropriate one. This is done by planning.
- Situation Calculus

Description of an action

- Possibility rule (precondition): a sentence that specifies when an action is possible.
- has(x, y, s) & apple(y) => Possible(eat(x, y), s)
- Effect rule (postcondition): a sentence that describes the consequences of an action.
- eat(x, y, s) => !exist(y, s+1)
- collect(x, y, s) => has(x, y, s+1)
- Frame problem: the collection of things / predicates that don't change.

Time and Events

- In time-based calculus the fluents are valid for particular moments in time instead of in a particular situation.
- A fluent is true at a particular point in time if it was initiated or caused by an event in the past, and was not terminated by another event in the meantime.
- Initiates(e, f, t), e-event, f-fluent, t-time.
- Terminates(e, f, t). Happens(e, t). Clipped(f, t, t2)
- T(f, t2) <=> ∃e, t (Happens(e, t) & Initiates(e, f, t) & (t < t2) & !Clipped(f, t, t2))
- Clipped(f, t, t2) <=> ∃ e, t1 (Happens(e, t1) & Terminates(e, f, t1) & (t < t1) & (t1 < t2))

Mental Processes

- Does an agent know that he knows something?
- We need to represent the fact that an agent believes something to be true or knows something to be true.
- We can make inferences about things an agent believes, things that the agent knows, and things that are true, but the combinations are not always possible.
- The inferences can be made only if the agent is a logical agent capable of reasoning.
- This is called modal logic.

Beliefs

- Believes(x, P(a)) x believes that P(a) is true.
- Inferences are not obvious.
- Believes(x, P(a)) & (a = b) does not imply that Believes(x, P(b)) is also true.
- We can apply reasoning within the system of beliefs for a logical agent:
- Believes(x, P(a)) & Believes(x, P(y) => Q(y)) can imply that Believes(x, Q(a)).

- We can't apply MP if the agent doesn't know or believe that P(y) => Q(y).

Knowledge and Belief

- Knows(x, P(a)) is similar to Believes, but the fact P(a) must be true.
- KnowsWhether(x, P(a)) means Knows(x, P(a)) | Knows(x, !P(a)).
- KnowsWhat(x, EyeColor(a)) means that ∃y Color(y) & Knows(x, (EyeColor(a) = y)).
- Believes(x, P(a)) & Knows(x, P(y)=>Q(y)) implies Believes(x, Q(a)).
- Knows(x, P(a)) & Believes (x, P(y)=>Q(y)) implies Believes(x, Q(a)).
- Knows(x, P(a)) & Knows(x, P(y)=>Q(y)) implies Knows(x, Q(a)).
- Fuzzy Logic
- First author: Lotfi Zadeh, University of California at Berkley, 1965.
- Def. Part of logic dealing with approximate reasoning, derived from the fuzzy set theory.
- A fuzzy set is one to which elements belong with a given percentage.
- In fuzzy logic the truth values belong to the interval [0,1] representing a percentage of truth.
- It is better suited to represent imprecise information /attributes from the real world and to make appropriate inferences.

Fuzzy Logic Expressions

- Classic logic: "Bill is a tall person."
- Tall(Bill) or Attribute(Bill, Tall) or
- in terms of sets, Bill belongs to TallPeopleSet

- Fuzzy logic: "Bill is 80% tall."
- Tall(Bill) = 0.8 or
- Bill belongs to TallPeopleSet by 80%
- We can also describe Bill as "quite tall" (80%) or "very tall"(95%) or "almost tall" (60%).
- It can apply to any comparable attributes:
- "Bill is taller than John."
- Tall(Bill) > Tall(John)

Hedges

- Hedges were introduced by Zadeh (MIT/IBM) and Lakoff (Berkley) in fuzzy logic and represent modifier to fuzzy sets:
- All-purpose: very, quite, extremely
- Truth values: quite true, mostly false
- Probabilities: likely, not very likely
- Quantifiers, most, several, few
- Possibilities, like almost impossible, quite possible.

Examples of Deductions

- This stream is narrow.
- This stream and that stream are approximately equal.
- Thus, that stream is more or less narrow.
- Most men are vain.
- Socrates is a man.
- Thus, Socrates is most likely vain.

FL Versus Probabilities

- Classic logic: "The pony is pretty."
- Either True or False.
- Fuzzy logic: "The pony is 90% pretty."
- This means that about 90% of the people looking at the pony will think it's pretty. We're talking about one

particular pony.

- Probabilities: "The pony has 90% chances to be pretty."
- This means that for the given population of ponies in our model, 90% of them are pretty. So 9 out of 10 ponies picked at random are expected to be pretty.

Operators

Zadeh operators:

- or - value(A | B) = max{value(A), value(B)}
- and - value(A & B) = min{value(A), value(B)}
- not - value(!A) = 1-value(A)
- Example: Annie is 90% tired and 75% cranky. Then Annie is
- 90% tired or cranky,
- 75% tired and cranky,
- 10% rested, and
- 25% is in a good mood.
- Other more complex metrics are also available.

Applications

- Automobile subsystems, such as ABS and cruise control
- Microcontrollers, microprocessors, robotics.
- Air conditioners, cameras, dishwashers, elevators.
- Search engines (web)
- The MASSIVE engine used in the Lord of the Rings films to create huge scale armies with random but somewhat orderly movements
- Digital image processing, edge detection.
- AI agents for games
- Language filters on message boards for filtering out offensive text.

Fuzzy Controller

- A controller using fuzzy values for variables and rules like
- IF variable IS value THEN action
- The rules are not mutually exclusive.
- Example: controller for an air conditioner:
- IF temperature IS very cold THEN stop fan
- IF temperature IS cold THEN turn down fan
- IF temperature IS normal THEN maintain level
- IF temperature IS hot THEN speed up fan

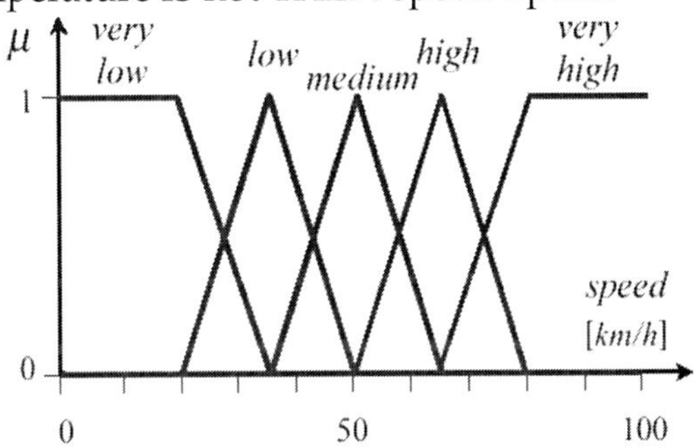

Example of Fuzzy Sets - car speed

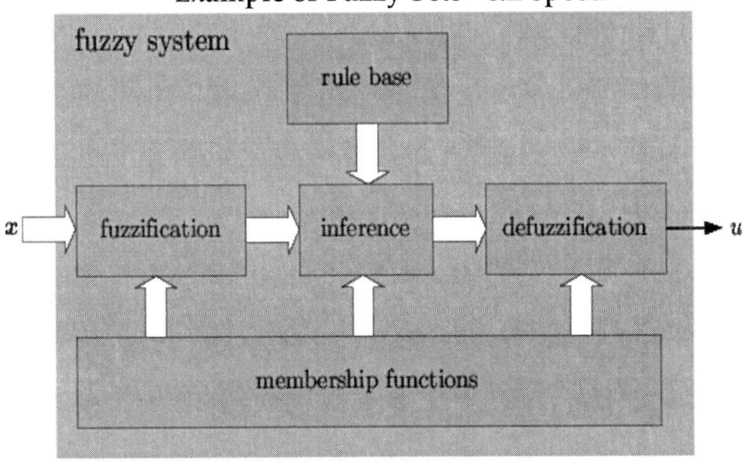

Fuzzy Controller Structure

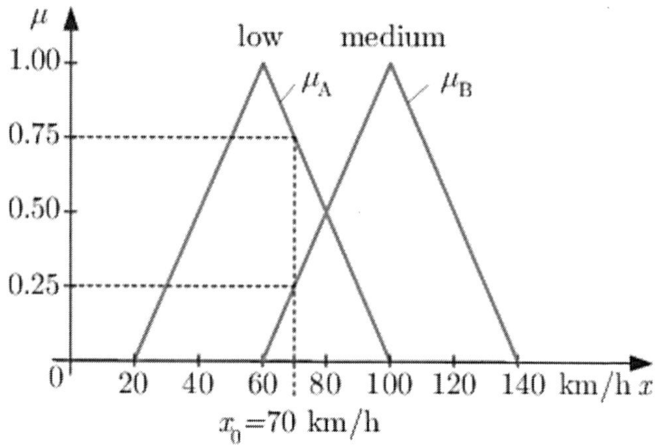

3.11 Reasoning Systems for Categories

The reasoning is the mental process of deriving logical conclusion and making predictions from available knowledge, facts, and beliefs. Or we can say, "**Reasoning is a way to infer facts from existing data.**" It is a general process of thinking rationally, to find valid conclusions.

In artificial intelligence, the reasoning is essential so that the machine can also think rationally as a human brain, and can perform like a human.

Types of Reasoning

In artificial intelligence, reasoning can be divided into the following categories:

1. Deductive reasoning
2. Inductive reasoning
3. Abductive reasoning
4. Common Sense Reasoning
5. Monotonic Reasoning
6. Non-monotonic Reasoning

1. Deductive reasoning

Deductive reasoning is deducing new information from logically related known information. It is the form of valid reasoning, which means the argument's conclusion must be true when the premises are true.

Deductive reasoning is a type of propositional logic in AI, and it requires various rules and facts. It is sometimes referred to as top-down reasoning, and contradictory to inductive reasoning.

In deductive reasoning, the truth of the premises guarantees the truth of the conclusion.

Deductive reasoning mostly starts from the general premises to the specific conclusion, which can be explained as below example.

Example:

Premise-1: All the human eats veggies Premise-2: Suresh is human. Conclusion: Suresh eats veggies.

The general process of deductive reasoning is given below:

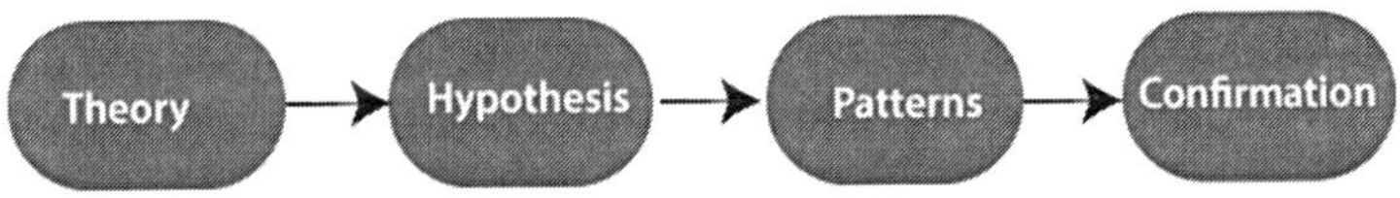

2. Inductive Reasoning

Inductive reasoning is a form of reasoning to arrive at a conclusion using limited sets of facts by the process of generalization. It starts with the series of specific facts or data and reaches to a general statement or conclusion.

Inductive reasoning is a type of propositional logic, which is also known as cause-effect reasoning or bottom-up reasoning.

In inductive reasoning, we use historical data or various premises to generate a generic rule, for which premises support the conclusion.

In inductive reasoning, premises provide probable supports to the conclusion, so the truth of premises does not guarantee the truth of the conclusion.

Example

Premise: All of the pigeons we have seen in the zoo are white. Conclusion: Therefore, we can expect all the pigeons to be white.

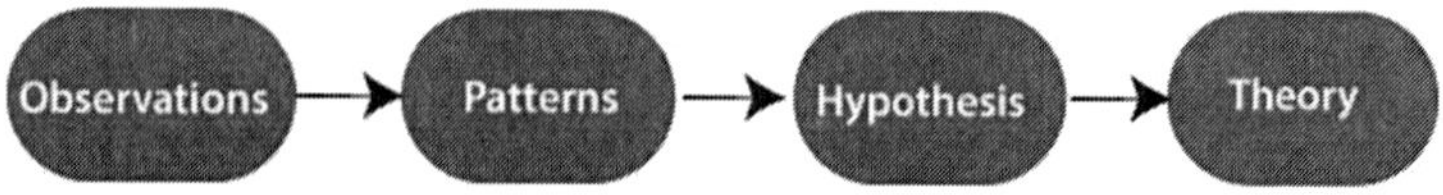

3. Abductive reasoning:

Abductive reasoning is a form of logical reasoning which starts with single or multiple observations then seeks to find the most likely explanation or conclusion for the observation.

Abductive reasoning is an extension of deductive reasoning, but in abductive reasoning, the premises do not guarantee the conclusion.

Example

Implication: Cricket ground is wet if it is raining
Axiom: Cricket ground is wet. Conclusion It is raining.

4. Common Sense Reasoning

Common sense reasoning is an informal form of reasoning, which can be gained through experiences.

Common Sense reasoning simulates the human ability to make presumptions about events which occurs on every day.

It relies on good judgment rather than exact logic and operates on **heuristic knowledge** and **heuristic rules**.

Example

1. One person can be at one place at a time.
2. If I put my hand in a fire, then it will burn.

The above two statements are the examples of common sense reasoning which a human mind can easily understand and assume.

5. Monotonic Reasoning

In monotonic reasoning, once the conclusion is taken, then it will remain the same even if we add some other information to existing information in our knowledge base. In monotonic reasoning, adding knowledge does not decrease the set of prepositions that can be derived.

To solve monotonic problems, we can derive the valid conclusion from the available facts only, and it will not be affected by new facts.

Monotonic reasoning is not useful for the real-time systems, as in real time, facts get changed, so we cannot use monotonic reasoning.

Monotonic reasoning is used in conventional reasoning systems, and a logic-based system is monotonic.

Any theorem proving is an example of monotonic reasoning.

Example

Earth revolves around the Sun

It is a true fact, and it cannot be changed even if we add another sentence in knowledge base like, "The moon revolves around the earth" Or "Earth is not round," etc.

Advantages of Monotonic Reasoning

- In monotonic reasoning, each old proof will always remain valid.
- If we deduce some facts from available facts, then it will remain valid for always.
- Disadvantages of Monotonic Reasoning:
- We cannot represent the real world scenarios using Monotonic reasoning.
- Hypothesis knowledge cannot be expressed with monotonic reasoning, which means facts should be true.

- Since we can only derive conclusions from the old proofs, so new knowledge from the real world cannot be added.

6. Non-monotonic Reasoning

In Non-monotonic reasoning, some conclusions may be invalidated if we add some more information to our knowledge base.

Logic will be said as non-monotonic if some conclusions can be invalidated by adding more knowledge into our knowledge base.

Non-monotonic reasoning deals with incomplete and uncertain models.

"Human perceptions for various things in daily life, "is a general example of non-monotonic reasoning.

Example: Let suppose the knowledge base contains the following knowledge:

- **Birds can fly**
- **Penguins cannot fly**
- **Pitty is a bird**

So from the above sentences, we can conclude that **Pitty can fly**.

However, if we add one another sentence into knowledge base "**Pitty is a penguin**", which concludes "**Pitty cannot fly**", so it invalidates the above conclusion.

Advantages of Non-monotonic reasoning

- For real-world systems such as Robot navigation, we can use non-monotonic reasoning.
- In Non-monotonic reasoning, we can choose probabilistic facts or can make assumptions.

Disadvantages of Non-monotonic Reasoning

- In non-monotonic reasoning, the old facts may be invalidated by adding new sentences.
- It cannot be used for theorem **proving**.

3.12 Reasoning with Default Information

When giving information, we don't want to enumerate all of the exceptions, even if we could think of them all. In default reasoning, we specify general knowledge and modularly add exceptions. The general knowledge is used for cases we don't know are exceptional. Classical logic is monotonic: If g logically follows from A, it also follows from any superset of A. Default reasoning is nonmonotonic: When we add that something is exceptional, we can't conclude what we could before.

Default reasoning can be modeled using H is normality assumptions F states what follows from the assumptions An explanation of g gives an argument for g. A reader of newsgroups may have a default: –Articles about AI are generally interesting‖.

H = {int ai},

where int ai means X is interesting if it is about AI. With facts:

interesting ← about ai ∧ int ai. about ai.

{int ai} is an explanation for interesting.

We can have exceptions to defaults:

false ← interesting ∧ uninteresting.

Suppose an article is about AI but is uninteresting:

interesting ← about ai ∧ int ai.

about ai.

uninteresting.

We cannot explain interesting even though everything we know about the previous we also know about this case

3-QUESTION-BANK

PART - A

1. Define FOL.

FOL is a first order logic. It is a representational language of knowledge which is powerful than propositional logic (i.e.) Boolean Logic. It is an expressive, declarative, compositional language.

2. Define a knowledge Base:

Knowledge base is the central component of knowledge base agent and it is described as a set of representations of facts about the world.

3. With an example, show objects, properties functions and relations. Example

EVIL KING JOHN BROTHER OF RICHARD RULED ENGLAND IN 1200‖

Objects : John, Richard, England, 1200

Relation : Ruled Properties : Evil, King Functions :

4. Define a Sentence?

Each individual representation of facts is called a sentence. The sentences are expressed in a language called as knowledge representation language.

5. Define an inference procedure

An inference procedure reports whether or not a sentence is entiled by knowledge base provided a knowledge base and a sentence .An inference procedure ‗i' can be described by the sentences that it can derive. If i can derive from knowledge base, we can write. KB --Alpha is derived from KB or i derives alpha from KB.

6. Define Ontological commitment.

The difference between propositional and first order logic is in the ontological commitment. It assumes about the nature of reality.

7. Define Epistemological commitment.

The logic that allows the possible states of knowledge with respect to each fact.

8. Define domain and domain elements.

The set of objects is called domain, sometimes these objects are referred as domain elements.

9. What are the three levels in describing knowledge based agent?

Logical level
Implementation level
Knowledge level or epistemological level

10. Define Syntax?

Syntax is the arrangement of words. Syntax of a knowledge describes the possible configurations that can constitute sentences. Syntax of the language describes how to make sentences.

11. Define Semantics

The semantics of the language defines the truth of each sentence with respect to each possible world. With this semantics, when a particular configuration exists within an agent, the agent believes the corresponding sentence.

12. Define Logic

Logic is one which consist of A formal system for describing states of affairs, consisting of

a) Syntax
b) Semantics. ii. Proof Theory – a set of rules for deducing the entailment of a set sentences.

13. What are the 3 types of symbol which is used to indicate objects, relations and functions?

b. Constant symbols for objects
c. Predicate symbols for relations c. Function symbols for functions

14. Define terms.

A term is a logical expression that refers to an object. We use 3 symbols to build a term.

15. Define Atomic sentence.

Atomic sentence is formed by both objects and relations.

Example

Brother (William, Richard) William is the brother of Richard.

16. Define Quantifier and it's types.

Quantifiers are used to express properties of entire collection of objects rather than representing the objects by name.

Types

a. Universal Quantifier
b. Existential Quantifier
c. Nested Quantifier.

17. What are the two we use to query and answer in knowledge base? ASK and TELL.

18. Define kinship domain.

The domain of family relationship is called kinship domain which consists of objects unary predicate, binary predicate, function, relation.

19. Define syntactic sugar.

Extension to the standard syntax (i.e.) procedure that does not change the semantics (i.e.) meaning is called syntactic sugar.

20. Define synchronic and diachronic sentence.

Sentences dealing with same time are called synchronic sentences. Sentences that allow reasoning –a cross time‖ are called diachronic sentence.

21. What are the 2 types of synchronic rules?

a. Diagnostic rules

b. Casual rules.

22. Define skolem constant.

The existential sentence says there is some object satisfying a condition, and the instantiation process is just giving a name to that object. That name must not belong to another object. The new name is called skolem constant.

23. What is truth Preserving

An inference algorithm that derives only entailed sentences is called sound or truth preserving

24. Define a Proof

A sequence of application of inference rules is called a proof. Finding proof is exactly finding solution to search problems. If the successor function is defined to generate all possible applications of inference rules then the search algorithms can be applied to find proofs.

25. Define a Complete inference procedure

An inference procedure is complete if it can derive all true conditions from a set of premises.

26. Define Interpretation

Interpretation specifies exactly which objects, relations and functions are referred to by the constant predicate, and function symbols.

27. Define Validity of a sentence

A sentence is valid or necessarily true if and only if it is true under all possible interpretation in all posssible world.

28. Define Satistiability of a sentence

A sentence is satisfiable if and only if there is some interpretation in some world for which it is true.

29. Define true sentence

A sentence is true under a particular interpretation if the state of affairs it represents is the case.

30. What are the basic Components of propositional logic?

Logical Constants (True, False)

Chapter 4
SOFTWARE AGENTS

4.1 What is an Agent?

An agent can be anything that perceiveits environment through sensors and act upon that environment through actuators. An Agent runs in the cycle of **perceiving, thinking**, and **acting**. An agent can be:

- **Human-Agent:** A human agent has eyes, ears, and other organs which work for sensors and hand, legs, vocal tract work for actuators.
- **Robotic Agent:** A robotic agent can have cameras, infrared range finder, NLP for sensors and various motors for actuators.
- **Software Agent:** Software agent can have keystrokes, file contents as sensory input and act on those inputs and display output on the screen.

Hence the world around us is full of agents such as thermostat, cellphone, camera, and even we are also agents before moving forward, we should first know about sensors, effectors, and actuators.

Sensor: Sensor is a device which detects the change in the environment and sends the information to other electronic devices. An agent observes its environment through sensors.

Actuators: Actuators are the component of machines that converts energy into motion. The actuators are only responsible for moving and controlling a system. An actuator can be an electric motor, gears, rails, etc.

Effectors: Effectors are the devices which affect the environment. Effectors can be legs, wheels, arms, fingers, wings, fins, and display screen.

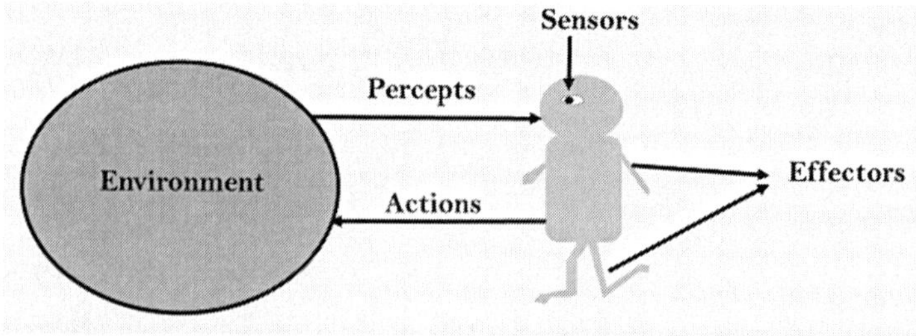

Intelligent Agents

An intelligent agent is an autonomous entity which act upon an environment using sensors and actuators for achieving goals. An intelligent agent may learn from the environment to achieve their goals. A thermostat is an example of an intelligent agent.

Following are the main four rules for an AI agent

- **Rule 1:** An AI agent must have the ability to perceive the environment.
- **Rule 2:** The observation must be used to make decisions.
- **Rule 3:** Decision should result in an action.
- **Rule 4:** The action taken by an AI agent must be a rational action.

Rational Agent

A rational agent is an agent which has clear preference, models uncertainty, and acts in a way to maximize its performance measure with all possible actions.

A rational agent is said to perform the right things. AI is about creating rational agents to use for game theory and decision theory for various real-world scenarios.

For an AI agent, the rational action is most important because in AI reinforcement learning algorithm, for each best possible action, agent gets the positive reward and for each wrong action, an agent gets a negative reward.

Rationality

The rationality of an agent is measured by its performance measure. Rationality can be judged on the basis of following points:

- Performance measure which defines the success criterion.
- Agent prior knowledge of its environment.
- Best possible actions that an agent can perform.
- The sequence of percepts.

Structure of an AI Agent

The task of AI is to design an agent program which implements the agent function. The structure of an intelligent agent is a combination of architecture and agent program. It can be viewed as:

Agent = Architecture + Agent program

Following are the main three terms involved in the structure of an AI agent:

Architecture: Architecture is machinery that an AI agent executes on.

Agent Function: Agent function is used to map a percept to an action.

$f:P^* \rightarrow A$

Agent program: Agent program is an implementation of agent function. An agent program executes on the physical architecture to produce function f.

PEAS Representation

PEAS is a type of model on which an AI agent works upon. When we define an AI agent or rational agent, then we can group its properties under PEAS representation model. It is made up of four words:

- **P:** Performance measure
- **E:** Environment
- **A:** Actuators
- **S:** Sensors

Here performance measure is the objective for the success of an agent's behavior.

PEAS for self-driving cars:

Performance: Safety, time, legal drive, comfort **Environment:** Roads, other vehicles, road signs, pedestrian **Actuators:** Steering, accelerator, brake, signal, horn **Sensors:** Camera, GPS, speedometer, odometer, accelerometer, sonar.

Example of Agents with their PEAS representation

Agent	Performance measure	Environment	Actuators	Sensors
1. Medical Diagnose	Healthy patient Minimized	Patient Hospital Staff	Tests Treatments	Keyboard (Entry of symptoms)
2. Vacuum Cleaner	Cleanness Efficiency Battery life Security	Room Table Wood floor Carpet Various obstacles	Wheels Brushes Vacuum Extractor	Camera Dirt detection sensor Cliff sensor Bump Sensor Infrared Wall Sensor
3. Part -picking Robot	Percentage of parts in correct bins.	Conveyor belt with parts, Bins	Jointed Arms Hand	Camera Joint angle sensors.

Agent Environment in AI

An environment is everything in the world which surrounds the agent, but it is not a part of an agent itself. An environment can be described as a situation in which an agent is present.

The environment is where agent lives, operate and provide the agent with something to sense and act upon it. An environment is mostly said to be non-feministic.

Features of Environment

As per Russell and Norvig, an environment can have various features from the point of view of an agent:

1. Fully observable vs Partially Observable
2. Static vs Dynamic
3. Discrete vs Continuous
4. Deterministic vs Stochastic
5. Single-agent vs Multi-agent
6. Episodic vs sequential
7. Known vs Unknown
8. Accessible vs Inaccessible

1. F ully observable vs Partially Observable:

- If an agent sensor can sense or access the complete state of an environment at each point of time then it is **a fully observable** environment, else it is **partially observable**.
- A fully observable environment is easy as there is no need to maintain the internal state to keep track history of the world.
- An agent with no sensors in all environments then such an environment is called as **unobservable**.

2. Deterministic vs Stochastic:

- If an agent's current state and selected action can completely determine the next state of the environment, then such environment is called a deterministic environment.
- A stochastic environment is random in nature and cannot be determined completely by an agent.
- In a deterministic, fully observable environment, agent does not need to worry about uncertainty.

3. Episodic vs Sequential:

- In an episodic environment, there is a series of one- shot actions, and only the current percept is required for the action.
- However, in Sequential environment, an agent requires memory of past actions to determine the next best actions.

4. Single-agent vs Multi-agent

- If only one agent is involved in an environment, and operating by itself then such an environment is called single agent environment.
- However, if multiple agents are operating in an environment, then such an environment is called a multi-agent environment.
- The agent design problems in the multi-agent environment are different from single agent environment.

5. Static vs Dynamic:

- If the environment can change itself while an agent is deliberating then such environment is called a dynamic environment else it is called a static environment.

- Static environments are easy to deal because an agent does not need to continue looking at the world while deciding for an action.
- However for dynamic environment, agents need to keep looking at the world at each action.
- Taxi driving is an example of a dynamic environment whereas Crossword puzzles are an example of a static environment.

6. Discrete vs Continuous:

- If in an environment there are a finite number of percepts and actions that can be performed within it, then such an environment is called a discrete environment else it is called continuous environment.
- A chess gamecomes under discrete environment as there is a finite number of moves that can be performed.
- A self-driving car is an example of a continuous environment.

7. Known vs Unknown

- Known and unknown are not actually a feature of an environment, but it is an agent's state of knowledge to perform an action.
- In a known environment, the results for all actions are known to the agent. While in unknown environment, agent needs to learn how it works in order to perform an action.
- It is quite possible that a known environment to be partially observable and an Unknown environment to be fully observable.

8. Accessible vs Inaccessible

- If an agent can obtain complete and accurate information about the state's environment, then such an environment is called an Accessible environment else it is called inaccessible.
- An empty room whose state can be defined by its temperature is an example of an accessible environment.
- Information about an event on earth is an example of Inaccessible environment.

4.2. Agent communication

Multiagent systems are distributed systems. Engineering a multiagent system means rigorously specifying the communications among the agents by way of interaction protocols. What makes specifying the protocols for agent interaction especially interesting and challenging is that agents are autonomous and heterogeneous entities. These properties of agents have profound implications on the nature of protocol specifications. As we shall see, protocols for multiagent systems turn out to be fundamentally different from those for other kinds of distributed systems such as computer networks and distributed databases.

We conceptualize all distributed systems in architectural terms—as consisting of components and connectors between the components. The components of the Internet are all nodes with IP addresses. The main connector is the Internet Protocol, which routes packets between the nodes. The components of the Web are the clients (such as browsers) and servers and the connector is the HTTP protocol.

The components in a distributed database are the client databases and the coordinator and a connector is the two- phase commit protocol. We can discern a pattern here: the connectors are nothing but the interaction protocols among the components. Further, we can associate protocols with the application it facilitates.

For example, the Internet Protocol facilitates routing; HTTP facilitates access to a distributed database of resources; and the two-phase commit protocol facilitates distributed transactions.

The same applies for multiagent systems except that the components are autonomous and heterogeneous agents, and applications are typically higher-level for example, auctions, banking, shipping, and so on. Each application would have its own set of requirements and therefore we would normally find different protocols for each application. Below, the term traditional distributed systems refers to non-multiagent distributed systems such as the Internet, the Web, and so on.

The importance of protocols is not lost upon industry. Communities of practice are increasingly interested in specifying standard protocols for their respective domains. RosettaNet [40] (e-business), TWIST [53] (foreign exchange transactions), GDSN [33] (supply chains), and HITSP [34] and HL7 [31] (health care) are just a few examples.

Our objectives in this chapter are to help the reader develop a clear sense of the conceptual underpinnings of agent communication and to help the reader learn to apply the concepts to the extent possible using available software. The chapter is broadly structured according to the following subobjectives.

Requirements for protocol specifications The inherently open nature of multiagent systems places certain requirements on protocol specifications. Meeting these requirements is the key to designing good protocols.

Protocol specification approaches There are many diverse approaches for specifying protocols. We evaluate some approaches widely practiced in software engineering and some historically significant ones from artificial intelligence.

We also study an approach that is particularly promising. Directions in agent communication research. The last fifteen years have seen some exciting developments in agent communication. However, many practical concerns remain to be addressed. We discuss these briefly

Autonomy and its Implications

Protocols are modular, potentially reusable specifications of interactions between two or more components. The interactions are specified in terms of the messages the components exchange. To promote reusability, a protocol is specified abstractly with reference to the roles that the interacting components may adopt.

A protocol is designed with a certain application in mind. An enactment refers to an execution of the protocol by the components.

Criteria for Evaluation

Communication has been studied in software engineering, distributed systems, and distributed artificial intelligence. Consequently, there are many approaches for specifying protocols. Later in the chapter, we discuss the major classes of approaches.

Let us now motivate broad criteria by which to evaluate each approach.

Software engineering Ideally, protocols should be specified in terms of highlevel abstractions that appeal to their stakeholders. In other words, protocol specifications should not be far removed from the expression of stakeholder requirements. Protocol specifications should be modifiable, easily understandable, and composable. Further, they should promote loose coupling among agents.

Flexibility Agents should be able to enact protocols flexibly. Flexibility is especially important in dynamic settings where agents may come and go, and exceptions and opportunities may arise. Ideally, protocol specifications should constrain agents no more than is necessary to ensure correctness, where correctness is understood in connection with the application domain of interest.

Compliance checking

An important standard of correctness is compliance.

Checking an agent's compliance with a protocol means determining if the agent is following the protocol. To make such a determination presupposes both that a protocol be precise and that its standard of correctness be based on information that is accessible to the agents involved.

4.3. Negotiation and Bargaining

Negotiation could be straightforward and conventional, such as bargaining over a price in a marketplace or determining on a meeting time or venue. It can also be complicated and extraordinary, such as involving international arguments and nuclear disarmament, which can impact the well-being of millions. According to Carnevale & Pruitt (1992, 1981), negotiation is defined as a procedure to resolve differing preferences among parties involving discussion with the aim of attaining agreement. Negotiation has also been defined as a process that involves decision making in a conflict between two parties while working together to achieve a satisfactory outcome . Negotiation exists in many professional subsets (e.g., business activities and legal proceedings) as well as private subsets (e.g., parenting and everyday life). It is also a pervasive activity and complex phenomenon appearing in many scientific fields, with a variety of very distinct approaches.

The negotiation process and successful completion is not an easy task depending on the complexity of a particular negotiation and the environment [1]. Successful negotiation, however, is critical to the negotiator. Therefore, it is crucial for negotiators to understand how to efficiently approach negotiations to reach a reasonable and mutually-beneficial settlement [3]. Unfortunately, it is often challenging for human negotiators to recognize and make required trade-offs to attain optimum results due to restricted information processing capacity and capability, cognitive biases, and social-emotional difficulties [4,5].

Artificial intelligence provides an opportunity and methods to transform from human-human negotiation to automated negotiation by implementing agent-based modeling. Such transformation is also important because of the fast growth of electronic markets (e-markets) requiring technologies to support human business decision-makers. Over past decades, a variety of negotiation models have been developed [6,7] and these can be applied to e-commerce and adapted for use by organizations.

Automated negotiation is interaction among a group of agents (software agents) in competing for interests and a willingness to collaborate to come up with a mutually-acceptable agreement regarding some belief, goal or plan. Therefore, comprehending the relationships between parties during the negotiation phase is a very significant problem in automated negotiation. Automated negotiation research has three primary research streams: negotiation protocols, negotiation objects, and agent decision-making models. Negotiation protocol refers to the flow of information between negotiating parties that is bound with the set of rules by which negotiating parties must abide. Negotiation object is the scope of the issue or item over which agreement must be reached (e.g., price). Last, agent decision-making is the decision- making object that participants employ with the negotiation procedure to reach their goals.

The formalization of automated negotiation started in the 1980s from the multi-agent community. Automated negotiation can significantly decrease transaction costs associated with humans and may increase the efficiency of negotiation time and settlement. It can also eliminate some of the uncommunicativeness and unnecessary human behavior that is typical during negotiation (e.g., personalities being offended).

Integrative Bargaining Model

The goal of integrative negotiation (also known as –non- zero-sum-game‖ or -win-win game‖) is to attain a mutually- beneficial agreement that maximizes settlement efficiency and fairness under suitable circumstances. Integrative approaches employ objective criteria to make a circumstance of mutual gain and emphasizes the significance of exchanging information among the negotiators. Walton and McKersie (1965) described the integrative bargaining model as a negotiation approach in which negotiators employ problem-solving behavior that refers to a state of desire for finding a solution to the problem to reach a definite goal. Problem-solving is generally recommended to achieve an integrative settlement. Negotiators attempt to redefine the problem, analyze the cause of settlement difficulties and explore a wide range of mutually-acceptable alternative solutions through maximum information sharing and disclosure of each party's needs and interests. The effectiveness of the problem-solving approach depends upon the presence of some psychological and information conditions: motivation, information and language, trust and a supportive climate. Motivation describes how parties must have the motivation to solve the problem and thus anticipate the problem as significant enough to address and discuss. Information and language state that those participating in the problem-solving process must have contact to appropriate information relevant and be authorized to use it. Meanwhile, the trust and support climate is marked by encouragement and freedom to behave spontaneously without fear of sanctions.

Negotiation Offer Strategy

Offer strategy is the action plan associated with the decision of negotiator to suggest an offer to the other party. To make negotiation successful, we suggested the offer strategy model that includes two features in negotiation: argumentation-based negotiation (as strategy) and multiple equivalent simultaneous offer (as tactics). The offer strategy model was based on the integrative approach and negotiation theory, which emphasizes the significance of exchanging information among negotiators . Meanwhile, multi-issue negotiation that contains package offers helps to reach a mutually acceptable (win-win) result.

Multiple Equivalent Simultaneous Offer (MESO)

The multiple equivalent simultaneous offer (MESO) tactics was suggested as an alternative to basic strategies such as sequential-single offer. The MESO technique lets the negotiator provide the other negotiator(s) with multiple offers that are mutually equal at each round. For instance, a software vendor can simultaneously offer three similar software packages to a client: $1 million software package with payment in 30 days, the same software package for $1.5 million with payment in 120 days, or an enhanced software package for $1.35 million with payment in 30 days. Negotiation theory and previous researches have proven that multi-issue negotiation including package offers is superior in reaching an agreement to single-issue negotiation. According to a research, users who highly value choice prefer being offered with multiple options to with single one and mutual benefits can be gained when both sides are using MESO. Furthermore, an experiment on human-to-human negotiation demonstrated that when the MESO technique is used, acceptance rate goes up, and the other party's satisfaction with the offer also increases.

Argumentation-Based Negotiation (ABN)

Argumentation theory is the interdisciplinary study of how deductions can be reached by logical reasoning. Logical reasoning is based on claims, if sound or not, and on premises. It includes the arts and science of civil debate, persuasion, dialogue and conversation. Arguments constitute the key part of real-life negotiations on personal matters (e.g., a fight between family members over which TV channel to watch), and also extend to business deals (e.g., a contract between the supplier and the retailer). In the context of negotiation, an argument is viewed as a piece of information that may permit an agent to justify its negotiation viewpoint and influence the other agent's negotiation viewpoint. For instance, in a trade union argument, an agent representing the workers' union might refuse an offer for a modified pension plan proposeded by the organization's management. As a response, the management agent might offer a different pension plan to persuade the workers' union agent to accept it as a win-win settlement. It is necessary to incorporate arguments into the negotiation model for the following two reasons:

a. argumentation is a tool in the negotiation process in which an agent gathers information and strategically discloses the information to adjust utility functions and to update beliefs of the other party;
b. in reality, agents frequently have limited (as opposed to zero or full) knowledge of all the situations around the negotiation, and argumentation is used by an agent to influence its opponent by strategically providing selected pieces of information to the opponent.

4.4. Argumentation among Agents

A verbal and social activity of reason aimed at increasing (or decreasing) the acceptability of a controversial standpoint for the listener or reader, by putting forward a constellation of propositions (i.e. arguments) intended to justify (or refute) the standpoint before a rational judge‖

What Is Argumentation Good for? According to a recent authoritative reference on argumentation theory, argumentation can be defined as follows:

Argumentation is a verbal and social activity of reason aimed at increasing (or decreasing) the acceptability of a controversial standpoint for the listener or reader, by putting forward a constellation of propositions intended to justify (or refute) the standpoint before a rational judge.

Let us decompose the elements of this definition that are most relevant to our discussion. First, the ultimate goal of argumentation is to resolve a ―controversial‖ standpoint; controversial in the sense that it is subject to both justification‖ or refutation‖ depending on the information available. This distinguishes argumentation from the classical deductive reasoning viewpoint, in which proofs for propositions cannot be contested. Moreover, the nature of the –standpoint‖ can vary. While the classical study of argumentation has focused mainly on propositional standpoints – i.e. things that are believed or known – there is no reason why the standpoint is confined to be propositional. A standpoint can, in principle, range from a proposition to believe, to a goal to try to achieve, to a value to try to promote. That is, argumentation can be used for theoretical reasoning (about what to believe) as well as practical reasoning (about what to do). Secondly, argumentation is an –activity of reason‖, emphasising that a particular process is to be followed in order to influence the acceptability of the controversial standpoint. This activity and the propositions put forward are to be evaluated by a –rational judge‖: a system that defines the reasonableness of these propositions according to some criteria. An important objective of argumentation theory is to identify such system of criteria.

In summary, argumentation can be seen as the principled interaction of different, potentially conflicting arguments, for the sake of arriving at a consistent conclusion. Perhaps the most crucial aspect of argumentation is the interaction between arguments. Argumentation can give us means for allowing an agent to reconcile conflicting information within itself, for reconciling its informational state with new perceptions from the environment, and for reconciling conflicting information between multiple agents through communication. It is for these reasons that argumentation has begun to receive great interest in the multi-agent systems community. In particular, argumentation lends itself naturally to two main sorts of problems encountered in MAS:

Forming and revising beliefs and decisions:

Argumentation provides means for forming beliefs and decisions on the basis of incomplete, conflicting or uncertain information. This is because argumentation provides a systematic means for resolving conflicts among different arguments and arriving at consistent, well-supported standpoints; - Rational interaction: Argumentation provides means for structuring dialogue between participants that have potentially conflicting viewpoints. In particular, argumentation provides a framework for ensuring that interaction respects certain principles (e.g. consistency of each participant's statements). In the next sections, we will discuss these applications in more detail and refer to some relevant literature. In particular, Section 3 deals with the topics of revising beliefs and making decisions, aspects that we can think of as being the concern of individual autonomous agents, while Section 4 deals with topics related to interagent communication and rational action, all aspects of argumentation that are decidedly multi-agent.

Argumentation for Reasoning in Autonomous Agents Argumentation is a general process for reasoning. An autonomous agent that has to reason about could weigh arguments for and against different options in order to arrive at a well-supported stance. In this section, we discuss two main applications of argumentation to autonomous agent reasoning. 3.1 Argumentation for Belief Revision One of the main challenges in specifying autonomous agents is the maintenance and updating of its beliefs in a dynamic environment. An agent may receive perceptual information that is inconsistent with its view of the world, in which case the agent needs to update its beliefs in order to maintain consistency. The major challenge of nonmonotonic reasoning formalisms [11] is to specify efficient ways to update beliefs. At the normative level, the AGM paradigm [29] specifies the rationality postulates that must be satisfied by an idealistic process of belief revision. On the operational level, formalisms for mechanising nonmonotonic reasoning include truth maintenance systems (TMS) [19], default logic [75] and circumscription [48]. Argumentation provides an alternative way to mechanise nonmonotonic reasoning. Argument-based frameworks view the problem of nonmonotonic reasoning as a process in which arguments for and against certain conclusions are constructed and compared. Nonmonotonicity arises from the fact that new premises may enable the construction of new arguments to support new beliefs, or stronger counterarguments against existing beliefs. As the number of premises grows, the set of arguments that can be constructed from those premises grows monotonically. However, because new arguments may overturn existing beliefs, the set of beliefs is nonmonotonic. Various argument- based frameworks for nonmonotonic reasoning have been proposed in the last 20 or so years. Some of the most notable are the following While the above-mentioned frameworks have developed into a solid and mature sub-field of AI, their incorporation into situated autonomous agent reasoning remains an opportunity to be pursued. In order to do so, an adequate representation of the environment is needed, and a mechanism for integrating perceptual information into the belief-update mechanism is also required. Moreover, situated agents are required to update their beliefs in a timely fashion in order to take appropriate action accordingly.

An autonomous agent does not only maintain a mental picture of its environment. The agent is faced with two additional tasks: the task of deliberation in which it decides what state of the world it wishes to achieve – namely its goal and the task of means-ends reasoning in which it forms a plan to achieve this goal. Argumentation is also potentially useful for tackling both these challenges. Recently, argumentation has been applied to deliberation. For example, argumentation has been used as a means for choosing among a set of conflicting desires and as a means for choosing between goals Another argumentbased framework for deliberation has been presented by Kakas and Moraitis. In this approach, arguments and preferences among them are used in order to generate goals based on a changing context. In addition, argumentation can be used to support standard BDI

Argumentation for Agent Communication

An inherent, almost defining, characteristic of multi-agent systems is that agents need to communicate in order to achieve their individual or collective aims. Argumentation theory has been an inspiration for studying and formalising various aspects of agent communication. Enhancing agent communication with argumentation allows agents to exchange arguments, to justify their stance, to provide reasons that defend their claims. This improved expressivity has many potential benefits, but it is often claimed that it should in particular: - make communication more efficient by allowing agents to reveal relevant pieces of information when it is required during a conversation; - allow for a verifiable semantics based on the agents' ability to justify their claims (and not on private mental states); and - make protocols more flexible, by replacing traditional protocol-based regulation by more sophisticated mechanics based on commitments. On the other hand, this improved expressivity comes with a price: it poses some serious challenges when it comes to designing autonomous agents that actually communicate by means of arguments, and makes more difficult: - the integration with agents' reasoning, which requires to precisely specify what agents should respond to others' agents on the basis of their internal state, but also on the basis of their goal (strategy); the validation of provable desirable properties of these protocols; - the communication between potentially heterogeneous agents, which should now share an argument interagent format.

4.1 Efficiency of Argumentation Until rather recently it was often claimed that argumentation could make communication more efficient, by allowing agents to reveal relevant pieces of information when it is needed during a conversation. Although the idea is intuitively appealing, there was little evidence to confirm this though. Indeed, argumentation may also involve both computational and communication overload, hence compensating the potential benefits induced by the exchange of reasons justifying agents' stances regarding an issue.

Without entering in the details of these experimental results, it is interesting to note that the efficiency of argumentation is very much dependent of the context, and that there can be no straightforward answer to the question ―Is it worth arguing?‖. For instance, Karunatillake and Jennings show that argumentation turns out to be effective when the the number of resources involved in the task allocation problem remains rather limited. Similarly, Bourgne et al. [10] observe that argumentation is especially required in those situations where buildings are rather ―open‖ (when there are fewer walls). This can be explained by the fact that there are more potential candidate hypotheses to the fire origin then, hence the need to exchange arguments to discriminate between those. In their multiagent learning experiment Onta˜n´on and Plaza emphasize in particular the influence of the amount of data that agents can individually access: as expected, argumentation is more beneficial when agents have only limited access to data. While these papers try to investigate mostly experimentally in what circumstances argumentation can be an efficient conflict resolution technique; there are more theoretical contributions to this issue. A recent paper by Rahwan et al. makes a first effort in this direction. In particular, the authors investigate a simple argumentation-based negotiation protocol in which agents exchange information about their underlying goals. It is shown that under certain conditions, exchanging such information enables agents to discover mutual goals and thus increases the likelihood of reaching deals. Other related work is that of which shows how the beliefs of two agents that engage in argumentation-based dialogue will converge over time

4.3 Integration of Argumentation and Reasoning We have seen that argumentation can serve both as a framework for implementing autonomous agent reasoning (e.g. about beliefs and actions) and as a means to structure communication among agents. As a result, argumentation can naturally provide a means for integrating communication with reasoning in a unified framework. To illustrate the above point, consider the following popular example by Parsons et al. [56]. The example concerns two home-improvement agents agent A1 trying to hang a painting, and another A2 trying to hang a mirror. A1 possesses a screw, a screw driver and a hammer, but needs a nail in addition to the hammer to hang the painting. On the other hand, A2 possesses a nail, and believes that to hang the mirror, it needs a hammer in addition to the nail. Now, consider the following dialogue (described here in natural language) between the two agents: A1: Can you please give me a nail? A2: Sorry, I need it for hanging a mirror. A1: But you can use a screw and a screw driver to hang the mirror! And if you ask me, I can provide you with these. A2: Really? I guess in that case, I do not need the nail. Here you go. A1: Thanks. At first, A2 was not willing to give away the nail because it needed it to achieve its goal. But after finding out the reason for rejection, A1 managed to persuade A2 to give away the nail by providing an alternative plan for achieving the latter's goal. We can use this example to highlight how argumentation-based techniques can provide a comprehensive set of features required for communication. Let us consider these in detail.

1. Reasoning and Plan ning: Argumentation can be used by each agent to form its beliefs about the environment, and to generate plans for achieving their goals. For example, agent A2 can use argument-based deliberation to arrive at the goal to acquire a nail. 2. Generating Utterances: Argumentation can be used to generate arguments for utterances and arguments. For example, after A1 requests a nail from A2, the latter builds an argument against giving away the nail by stating that it needs the nail to achieve one of its own goals (namely, hanging the mirror). This information can be used again by A2 to generate a counterargument for why A2 does not need the nail. 3. Evaluating incoming communication: Argumentation- based belief revision can be used to evaluate incoming communication. For example, when A2 received the argument from A1, it had to evaluate that argument to make sure it is sensible. A2 would not have accepted A1's argument if the former did not believe the latter actually possesses a screw and screw driver. 4. Communication Structuring: The whole dialogue can be structured through argumentation-based protocols, based on dialogue-games, which may themselves be based on certain argumentation schemes for reasoning about resources and plans.

Argument Interchange Format

One major barrier to the development and practical deployment of argumentation systems is the lack of a shared, agreed notation for an –interchange format‖ for arguments and argumentation. Such a format is necessary if agents are to be able to exchange argumentative statements in open systems. The recently proposed Argument Interchange Format (AIF) is intended to fill this gap, providing an approach to the representation and exchange of data between various argumentation tools and agent-based applications. It represents a consensus –abstract model‖ established by researchers across the fields of argumentation, artificial intelligence and multi-agent systems. The core AIF ontology is specified in a way that it can be extended to capture a variety of argumentation formalisms and schemes. One such extension, in the context of Semantic Web applications, deals with Walton's theoretical model of argument schemes.

Concluding Remarks Argumentation theory has been concerned with the study of rational human reasoning and dialogue for millennia. It is therefore an ideal resource for techniques, results and intuitions for problems in multi-agent reasoning and communication, and it is no surprise that formal models of argumentation are becoming an increasingly popular subject within research on multi-agent systems. This chapter has presented a brief survey of a section of the work on argumentation in multi-agent systems, a section that encompasses the work that, in the opinion of the authors, is currently the most interesting of the work in the field. In short, in our view, the basic tools and methods have been established we have well founded argumentation systems, and we have in dialogue games a means of structuring interactions between agents. What we need to do is to work with these tools in three directions. First, we need to integrate them into the reasoning processes of agents. For example, we need to decide how what an agent knows informs what it chooses to say in an interaction, and, conversely, what is said in an interaction informs what an agent knows. Second, we need to understand better how to design argumentation-based agent interactions so that they achieve the things that we want — we don't just need theoretical results that tell us how specific protocols work, but we need a theory that tells us how all protocols work. Third, we need to be able to show the effectiveness of argumentation-based agent interactions. In the end, however attractive the theory, if argumentation-based approaches are not more effective than other approaches to creating interactions between agents, then work on them is work wasted. As the paper surveyed above, and the work described in the contributions to this volume, show, as a community we are taking some steps in these three important directions.

4.5 Trust and Reputation in Multi-agent systems.

A multi-agent system is one in which many agents run concurrently, communicating amongst themselves and working toward either individual goals or a common objective. Jennings suggested that networked computer systems such as the Grid and Semantic Web can be viewed as multi-agent systems, as individual components in each system are both autonomous and flexible in their actions. Trust and reputation are widely described as central to effective interactions between agents; they are defined as expectation of competence and willingness to perform a given task. They can be quantified using various different algorithms involving past experience, some giving relevance to opinions held by an agent's neighbours, and others using only a single agent's own previous experience. Trust is most beneficial when agents have the opportunity to have a large amount of interactions with one another.

Basic trust is how intrinsically trusting an agent's disposition is, regardless of what situation it's in and which agent it is interacting with. The basic trust value is calculated from all accumulated experiences; for example, if an agent encounters many bad experiences, it is less likely to be trusting than an agent which has only encountered good experiences. General trust is how much trust one agent has for another, regardless of any situation either agent might be in. This is calculated on a per-agent basis; i.e. each agent will have a list of other agents with their corresponding trust values. Situational trust is how much an agent trusts another agent in a given situation. The calculation of situational trust values takes into account the utility of the situation, how important it is and also the general trust value for that particular pair of agents. Marsh states that there are three methods to calculate situational trust: optimistically (taking the maximum trust value), realistically (taking the mean trust value) and pessimistically (taking the minimum trust value). Once these three trust values have been obtained, the agent can use them to decide whether or not it is worth it's while co-operating with a certain agent. In making this decision, the agent will also take into account the importance of the action to be performed, any risks involved in the current situation, and how competent the target agent is perceived to be at the task required. Mar the trust metric employed by the website eBay1 , despite not being outlined in a specific paper, is an important one to the field of trust and reputation because it shows a real-life application of it, and is discussed by Huynh et al. The author discusses how eBay utilises a centralised reputation mechanism; all transaction feedback ratings are stored in a central queriable repository. Users can discover each other's feedback ratings by viewing appropriate user profiles on the eBay website. These profiles contain all collected feedback ratings for the user, as well as an overall feedback rating: a percentage relating to the ratio of good and bad ratings for that user. eBay users have to employ a specific kind of trust, which is called service provision trust. Service provision trust is the trust that a service provider (a seller, in eBay terms) is competent

4-QUESTION BANK
PART - A

1. Define the term utility?

The term utility is used in the sense of "the quality of being useful.", utility of a state is relative to the agents, whose preferences the utility function is supposed to represent.

2. What is the need for probability theory in uncertainty?

Probability provides the way of summarizing the uncertainty that comes from our laziness and ignorance. Probability statements do not have quite the same kind of semantics known as evidences.

3. What is the need for utility theory in uncertainty?

Utility theory says that every state has a degree of usefulness, or utility to in agent, and that the agent will prefer states with higher utility. The use utility theory to represent and reason with preferences.

4. What is called as principle of maximum expected utility?

The basic idea is that an agent is rational if and only if it chooses the action that yields the highest expected utility, averaged over all the possible outcomes of the action. This is known as MEU.

5. What Is Called As Decision Theory?

Preferences As Expressed by Utilities Are Combined with Probabilities in the General Theory of Rational Decisions Called Decision **Theory**.

Decision Theory = Probability Theory + Utility Theory.

6. Define Prior Probability?

p(a) for the Unconditional or Prior Probability Is That the Proposition A is True. It is important to remember that p(a) can only be used when there is no other information.

7. Define conditional probability?

Once the agents has obtained some evidence concerning the previously unknown propositions making up the domain conditional or posterior probabilities with the notation p(A/B) is used. This is important that p(A/B) can only be used when all be is known.

8. Define probability distribution:

Eg. P(weather) = (0.7,0.2,0.08,0.02). This type of notations simplifies many equations.

9. What is an atomic event?

An atomic event is an assignment of particular values to all variables, in other words, the complete specifications of the state of domain.

10. Define joint probability distribution

This completely specifies an agent's probability assignments to all propositions in the domain. The joint probability distribution p(x1,x2,--------xn) assigns probabilities to all possible atomic events; Where X1,X2------Xn 10 –variables.

11. What are the ways in which one can understand the semantics of a belief network?

There are two ways to see the network as a representation of the joint probability distribution to view it as an encoding of collection of conditional independence statements.

12. What is the basic task of a probabilistic inference?

The basic task is to reason in terms of prior probabilities of conjunctions, but for the most part, we will use conditional probabilities as a vehicle for probabilistic inference.

13. What are called as Poly trees?

The algorithm that works only on singly connected networks known as Poly trees. Here at most one undirected path between any two nodes is present.

14. Define casual support

E+X is the casual support for X- the evidence variables "above" X that are connected to X through its parent.

15. Define evidential support

E-X is the evidential support for X- the evidence variables "below" X that are connected to X through its children.

16. What is called as multiple connected graph?

A multiple connected graph is one in which two nodes are connected by more than one path.

17. What is the purpose of learning?

The idea behind learning is that percepts should be used not only for acting but also for improving the agent's ability to act in the future.

18. What are issues in learning element?

a. Component
b. Feedback
c. Representation

19. What are the types of machine learning?
    a. Supervised Learning
    b. Unsupervised Learning c. Reinforcement Learning

20. Differentiate Supervised and Unsupervised learning.

Supervised Learning	Unsupervised Learning
It involves learning a function from examples of its inputs And Outputs	It involves learning patterns in the input when no specific output values are supplied.
Example: Applying Brake on the wet road, we can even skip on the	Example: Day by day agent will learn about ―Good traffic days‖ and ―Bad traffic

Chapter 5
APPLICATIONS

5.1 AI Application

Artificial Intelligence has various applications in today's society. It is becoming essential for today's time because it can solve complex problems with an efficient way in multiple industries, such as Healthcare, entertainment, finance, education, etc. AI is making our daily life more comfortable and fast.

Following are some sectors which have the application of Artificial Intelligence:

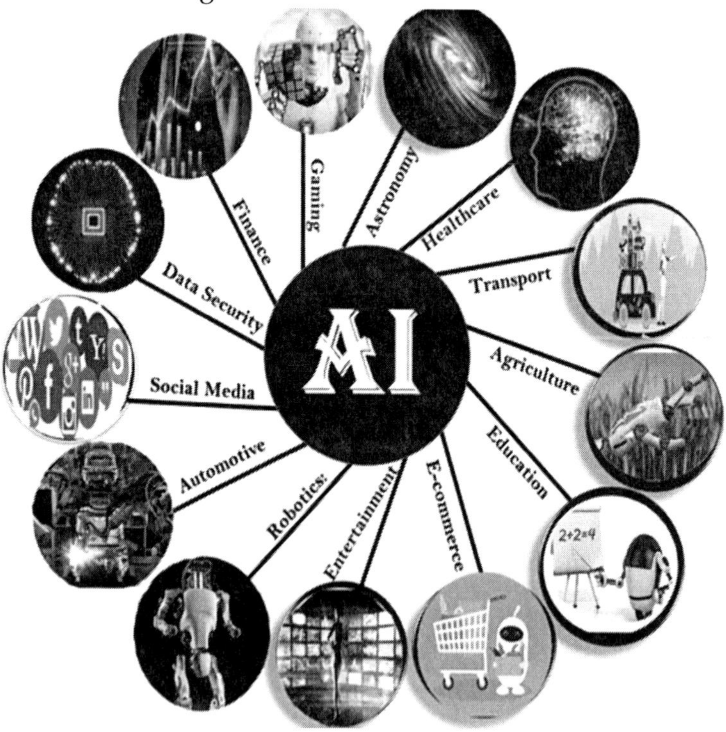

1. AI in Astronomy

- Artificial Intelligence can be very useful to solve complex universe problems. AI technology can be helpful for understanding the universe such as how it works, origin, etc.

2. AI in Healthcare

- In the last, five to ten years, AI becoming more advantageous for the healthcare industry and going to have a significant impact on this industry.
- Healthcare Industries are applying AI to make a better and faster diagnosis than humans. AI can help doctors with diagnoses and can inform when patients are worsening so that medical help can reach to the patient before hospitalization.

3. AI in Gaming

- AI can be used for gaming purpose. The AI machines can play strategic games like chess, where the machine needs to think of a large number of possible places.

4. AI in Finance

- AI and finance industries are the best matches for each other. The finance industry is implementing automation, chatbot, adaptive intelligence, algorithm trading, and machine learning into financial processes.

5. AI in Data Security

- The security of data is crucial for every company and cyber-attacks are growing very rapidly in the digital world. AI can be used to make your data more safe and secure. Some examples such as AEG bot, AI2 Platform, are used to determine software bug and cyber-attacks in a better way.

6. AI in Social Media

- Social Media sites such as Facebook, Twitter, and Snapchat contain billions of user profiles, which need to be stored and managed in a very efficient way. AI can organize and manage massive amounts of data. AI can analyze lots of data to identify the latest trends, hashtag, and requirement of different users.

7. AI in Travel & Transport

- AI is becoming highly demanding for travel industries. AI is capable of doing various travel related works such as from making travel arrangement to suggesting the hotels, flights, and best routes to the customers. Travel industries are using AI-powered chatbots which can make human-like interaction with customers for better and fast response.

8. AI in Automotive Industry

- Some Automotive industries are using AI to provide virtual assistant to their user for better performance. Such as Tesla has introduced TeslaBot, an intelligent virtual assistant.
- Various Industries are currently working for developing self-driven cars which can make your journey more safe and secure.

9. AI in Robotics

- Artificial Intelligence has a remarkable role in Robotics.

Usually, general robots are programmed such that they can perform some repetitive task, but with the help of AI, we can create intelligent robots which can perform tasks with their own experiences without pre- programmed.

- Humanoid Robots are best examples for AI in robotics, recently the intelligent Humanoid robot named as Erica and Sophia has been developed which can talk and behave like humans.

10. AI in Entertainment

- We are currently using some AI based applications in our daily life with some entertainment services such as Netflix or Amazon. With the help of ML/AI algorithms, these services show the recommendations for programs or shows.

11. AI in Agriculture

- Agriculture is an area which requires various resources, labor, money, and time for best result. Now a day's agriculture is becoming digital, and AI is emerging in this field. Agriculture is applying AI as agriculture robotics, solid and crop monitoring, predictive analysis. AI in agriculture can be very helpful for farmers.

12. AI in E-commerce

- AI is providing a competitive edge to the e-commerce industry, and it is becoming more demanding in the e-commerce business. AI is helping shoppers to discover associated products with recommended size, color, or even brand.

13. AI in education

- AI can automate grading so that the tutor can have more time to teach. AI chatbot can communicate with students as a teaching assistant.

- AI in the future can be work as a personal virtual tutor for students, which will be accessible easily at any time and any place.

5.2 Language Models

Language modeling (LM) is the use of various statistical and probabilistic techniques to determine the probability of a given sequence of words occurring in a sentence. Language models analyze bodies of text data to provide a basis for their word predictions. They are used in natural language processing (NLP) applications, particularly ones that generate text as an output. Some of these applications include , machine translation and question answering.

How language modeling works

Language models determine word probability by analyzing text data. They interpret this data by feeding it through an algorithm that establishes rules for context in natural language. Then, the model applies these rules in language tasks to accurately predict or produce new sentences. The model essentially learns the features and characteristics of basic language and uses those features to understand new phrases.

There are several different probabilistic approaches to modeling language, which vary depending on the purpose of the language model. From a technical perspective, the various types differ by the amount of text data they analyze and the math they use to analyze it. For example, a language model designed to generate sentences for an automated Twitter bot may use different math and analyze text data in a different way than a language model designed for determining the likelihood of a search query.

Some common statistical language modeling types are

- **N-gram.** N-grams are a relatively simple approach to language models. They create a probability distribution for a sequence of *n* The *n* can be any number, and defines the size of the "gram", or sequence of words being assigned a probability. For example, if *n* = 5, a gram might look like this: "can you please call me." The model then assigns probabilities using sequences of *n* size. Basically, *n* can be thought of as the amount of context the model is told to consider. Some types of n-grams are unigrams, bigrams, trigrams and so on.
- **Unigram.** The unigram is the simplest type of language model. It doesn't look at any conditioning context in its calculations. It evaluates each word or term independently. Unigram models commonly handle language processing tasks such as information retrieval. The unigram is the foundation of a more specific model variant called the query likelihood model, which uses information retrieval to examine a pool of documents and match the most relevant one to a specific query.
- **Bidirectional.** Unlike n-gram models, which analyze text in one direction (backwards), bidirectional models analyze text in both directions, backwards and forwards. These models can predict any word in a sentence or body of text by using every other word in the text. Examining text bidirectionally increases result accuracy. This type is often utilized in machine learning and speech generation applications. For example, Google uses a bidirectional model to process search queries.
- **Exponential.** Also known as maximum entropy models, this type is more complex than n-grams. Simply put, the model evaluates text using an equation that combines feature functions and n-grams. Basically, this type

specifies features and parameters of the desired results, and unlike n-grams, leaves analysis parameters more ambiguous -- it doesn't specify individual gram sizes, for example. The model is based on the principle of entropy, which states that the probability distribution with the most entropy is the best choice. In other words, the model with the most chaos, and least room for assumptions, is the most accurate. Exponential models are designed maximize cross entropy, which minimizes the amount statistical assumptions that can be made. This enables users to better trust the results they get from these models.

- **Continuous space.** This type of model represents words as a non-linear combination of weights in a neural network. The process of assigning a weight to a word is also known as word embedding. This type becomes especially useful as data sets get increasingly large, because larger datasets often include more unique words. The presence of a lot of unique or rarely used words can cause problems for linear model like an n-gram. This is because the amount of possible word sequences increases, and the patterns that inform results become weaker. By weighting words in a non- linear, distributed way, this model can "learn" to approximate words and therefore not be misled by any unknown values. Its "understanding" of a given word is not as tightly tethered to the immediate surrounding words as it is in n-gram models.

The models listed above are more general statistical approaches from which more specific variant language models are derived. For example, as mentioned in the n-gram description, the query likelihood model is a more specific or specialized model that uses the n-gram approach. Model types may be used in conjunction with one another.

The models listed also vary significantly in complexity.

Broadly speaking, more complex language models are better at NLP tasks, because language itself is extremely complex and always evolving. Therefore, an exponential model or continuous space model might be better than an n-gram for NLP tasks, because they are designed to account for ambiguity and variation in language.

A good language model should also be able to process long-term dependencies, handling words that may derive their meaning from other words that occur in far-away, disparate parts of the text. An LM should be able to understand when a word is referencing another word from a long distance, as opposed to always relying on proximal words within a certain fixed history. This requires a more complex model.

Importance of language modeling

Language modeling is crucial in modern NLP applications. It is the reason that machines can understand qualitative information. Each language model type, in one way or another, turns qualitative information into quantitative information. This allows people to communicate with machines as they do with each other to a limited extent.

It is used directly in a variety of industries including tech, finance, healthcare, transportation, legal, military and government. Additionally, it's likely most people reading this have interacted with a language model in some way at some point in the day, whether it be through Google search, an autocomplete text function or engaging with a voice assistant.

The roots of language modeling as it exists today can be traced back to 1948. That year, Claude Shannon published a paper titled "A Mathematical Theory of Communication." In it, he detailed the use of a stochastic model called the Markov chain to create a statistical model for the sequences of letters in English text. This paper had a large impact on the telecommunications industry, laid the groundwork for information theory and language modeling. The Markov model is still used today, and n-grams specifically are tied very closely to the concept.

Uses and examples of language modeling

Language models are the backbone of natural language processing (NLP). Below are some NLP tasks that use language modeling, what they mean, and some applications of those tasks:

- Speech recognition -- involves a machine being able to process speech audio. This is commonly used by voice assistants like Siri and Alexa.
- Machine translation -- involves the translation of one language to another by a machine. Google Translate and Microsoft Translator are two programs that do this. SDL Government is another, which is used to translate foreign social media feeds in real time for the U.S. government.
- Parts-of-speech tagging -- involves the markup and categorization of words by certain grammatical characteristics. This is utilized in the study of linguistics, first and perhaps most famously in the study of the Brown Corpus, a body of composed of random English prose that was designed to be studied by computers. This corpus has been used to train several important language models, including one used by Google to improve search quality.
- Parsing -- involves analysis of any string of data or sentence that conforms to formal grammar and syntax rules. In language modeling, this may take the form of sentence diagrams that depict each word's relationship to

the others. Spell checking applications use language modeling and parsing.

- Sentiment analysis -- involves determining the sentiment behind a given phrase. Specifically, it can be used to understand opinions and attitudes expressed in a text. Businesses can use this to analyze product reviews or general posts about their product, as well as analyze internal data like employee surveys and customer support chats. Some services that provide sentiment analysis tools are Repustate and Hubspot's ServiceHub. Google's NLP tool -- called Bidirectional Encoder Representations from Transformers (BERT) -- is also used for sentiment analysis.
- Optical character recognition -- involves the use of a machine to convert images of text into machine encoded text. The image may be a scanned document or document photo, or a photo with text somewhere in it -- on a sign, for example. It is often used in data entry when processing old paper records that need to be digitized. In can also be used to analyze and identify handwriting samples.

1.3 Information Retrieval- Information Extraction

Information retrieval -- involves searching in a document for information, searching for documents in general, and searching for metadata that corresponds to a document. Web browsers are the most common information retrieval applications. Information retrieval is based on a query - you specify what information you need and it is returned in human understandable form.

Information extraction is about structuring unstructured information - given some sources all of the (relevant) information is structured in a form that will be easy for processing. This will not necessary be in human understandable form - it can be only for use of computer programs.

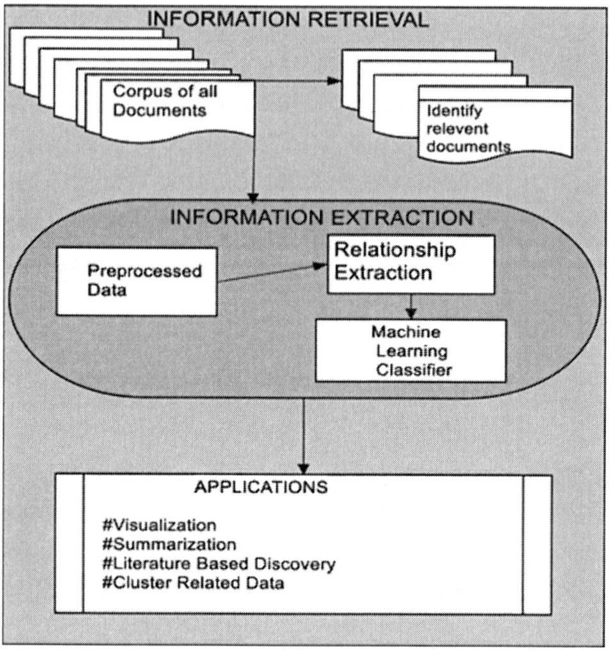

Information Extraction is not Information Retrieval: Information Extraction differs from traditional techniques in that it does not recover from a collection a subset of documents which are hopefully relevant to a query, based on key-word searching (perhaps augmented by a thesaurus).

Instead, the goal is to extract from the documents (which may be in a variety of languages) salient facts about prespecified types of events, entities or relationships. These facts are then usually entered automatically into a database, which may then be used to analyse the data for trends, to give a natural language summary, or simply to serve for on-line access.

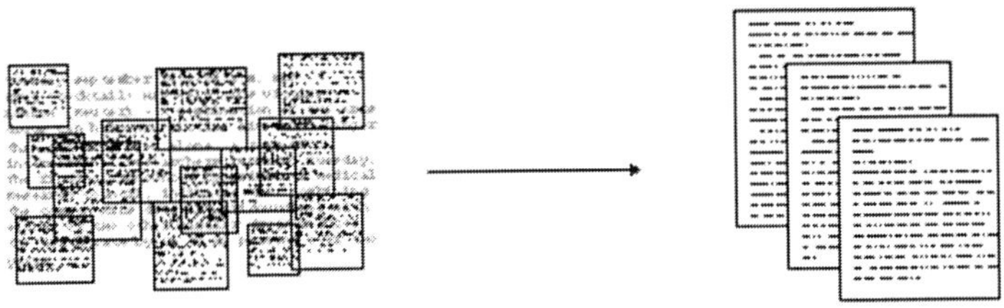

Information Retrieval gets sets of relevant documents

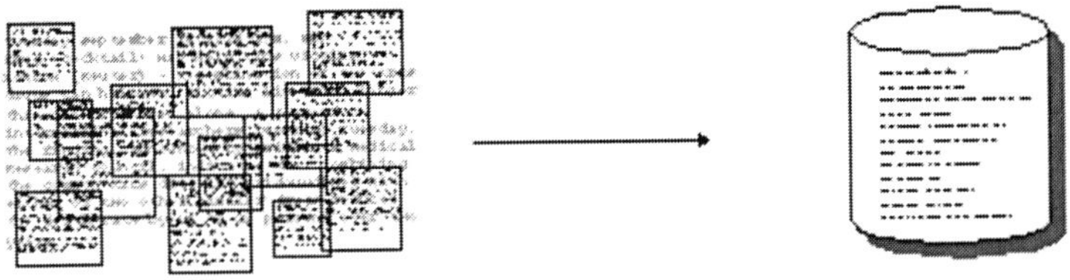

Information Extraction gets facts out of documents

From a modeling standpoint, information retrieval is a deep field predicated on several disciplines, including statistics, math, linguistics, artificial intelligence and now data science. In practice, these models are applied against text within corpora to discover patterns in the data. Not only do the IR models overlap in their usage, they can "partner" with other models such as k-means or k-nearest neighbor models, then other models can be applied from the vantage point of computational linguistics such as LDA/LDI and topic modeling Then, the end game is some sort of information visualization of this discovery--after ranking, clustering and aggregating work. Information Retrieval may appear to be a cryptic discipline, but serious effort, which is greatly appreciated, is going into opening up the area for deeper understanding of each model, and the interaction between models. I cite "Synthesis Lectures on Information Concepts, Retrieval, and Services" Series as the best place to delve into a foundation for IR.

While I don't entirely separate IR and Information Extraction, perhaps a subset of IE, concept level extraction, does apply IR patterns along with AI-based inferencing rules to extract related ontologies. The graphical nature of these relations are being enhanced with ontology modeling in OWL and RDF, and with graph databases, which allow for a less strict or rigorous set of relationship modeling, and allow for more relationships to surface, rather than being controlled per se. The ability to grow information extraction dynamically keeps its "discipline" strongly interesting to researchers.

Both IR and IE play out in our own significant "entities of the moment"--some have called "dynamic ontologies"--some being Palantir-- we need the patterns, models, simulations and visualizations of those significant entities to do business in the face of morphing new sources of information and changing of existing information. The conceptual, relational, definitional, pattern and ontological modeling have to be flexible and their visualizations the same. The heavy lifting of AI engines such as Watson in the information extraction and inferencing fields has cast a spotlight on the IE and frankly IR fields. Also the ubiquity of natural language processing and machine learning are calling attention to IR and IE models and engines. The impact of IR models on search and SEO, and on semantic web modeling is one of those "watching a change agent change due to impacts back from what it impacted" events--somewhere in harmonics and relativity theory. IR and IE are expanding based on what they're impacting.

Information retrieval is about returning the information that is relevant for a specific query or field of interest. Note that this information could also be in the form of general documents, sure enough search engines are a notable example of such task. I would say that the most important entities recognizable for information retrieval are the initial set of documents/information and the query that specify "what to search for".

On the other hand **information extraction** is more about extracting (or inferring) general knowledge (or relations) from a set of documents or information. Note that here all the content of the documents could be considered as a whole corpus of data from which extract the knowledge. Of course also for this case you can somehow specify what do you want to extract, but it is more about **properties/relations** than specific subjects/topics. Properties are more domain-specific, while generally relations cover more generic scenarios.

Information retrieval is about returning the information that is relevant for a specific query or field of interest. Note that this information could also be in the form of general documents, sure enough search engines are a notable example of such task. I would say that the most important entities recognizable for information retrieval are the initial set of documents/information and the query that specify "what to search for".

On the other hand **information extraction** is more about extracting (or inferring) general knowledge (or relations) from a set of documents or information. Note that here all the content of the documents could be considered as a whole corpus of data from which extract the knowledge. Of course also for this case you can somehow specify what do you want to extract, but it is more about **properties/relations** than specific subjects/topics. Properties are more domain-specific, while generally relations cover more generic scenarios.

Again, with search engines you're asking to get the sites that are most likely to contain information about that specific subject. This is an example of **information retrieval**.

For **information extraction** you could instead, for example, ask to extract all the names of cities, or e-mail addresses, that appear in a corpus of documents. You could even go much more generic, asking simply to extract knowledge. As you can see this is really generic, but it can be accomplish, for example, by obtaining triplets of the form subject-action-object for each valid sentence of a text (this is best suited for natural language texts).

1.4 Natural Language Processing

Natural Language Processing (NLP) refers to AI method of communicating with an intelligent systems using a natural language such as English.

Processing of Natural Language is required when you want an intelligent system like robot to perform as per your instructions, when you want to hear decision from a dialogue based clinical expert system, etc.

The field of NLP involves making computers to perform useful tasks with the natural languages humans use. The input and output of an NLP system can be –

- Speech
- Written Text

Components of NLP

There are two components of NLP as given – Natural Language Understanding (NLU) Understanding involves the following tasks –

- Mapping the given input in natural language into useful representations.
- Analyzing different aspects of the language.

Natural Language Generation (NLG)

It is the process of producing meaningful phrases and sentences in the form of natural language from some internal representation.

It involves −

- **Text planning** − It includes retrieving the relevant content from knowledge base.
- **Sentence planning** − It includes choosing required words, forming meaningful phrases, setting tone of the sentence.
- **Text Realization** − It is mapping sentence plan into sentence structure.

The NLU is harder than NLG. Difficulties in NLU NL has an extremely rich form and structure. It is very ambiguous. There can be different levels of ambiguity

- **Lexical ambiguity** − It is at very primitive level such as word-level.
- For example, treating the word ―board‖ as noun or verb?
- **Syntax Level ambiguity** − A sentence can be parsed in different ways.
- For example, ―He lifted the beetle with red cap.‖ − Did he use cap to lift the beetle or he lifted a beetle that had red cap?
- **Referential ambiguity** − Referring to something using pronouns. For example, Rima went to Gauri. She said, am tired.‖ − Exactly who is tired?
- One input can mean different meanings.
- Many inputs can mean the same thing.

NLP Terminology

- **Phonology** – It is study of organizing sound systematically.
- **Morphology** – It is a study of construction of words
- from primitive meaningful units.
- **Morpheme** – It is primitive unit of meaning in a
- language.
- **Syntax** – It refers to arranging words to make a sentence. It also involves determining the structural role of words in the sentence and in phrases.
- **Semantics** – It is concerned with the meaning of words and how to combine words into meaningful phrases and sentences.
- **Pragmatics** – It deals with using and understanding sentences in different situations and how the interpretation of the sentence is affected.
- **Discourse** – It deals with how the immediately preceding sentence can affect the interpretation of the next sentence.
- **World Knowledge** – It includes the general knowledge about the world.

Steps in NLP

There are general five steps –

- **Lexical Analysis** – It involves identifying and analyzing the structure of words. Lexicon of a language means the collection of words and phrases in a language. Lexical analysis is dividing the whole chunk of txt into paragraphs, sentences, and words.
- **Syntactic Analysis (Parsing)** – It involves analysis of words in the sentence for grammar and arranging words in a manner that shows the relationship among the

words. The sentence such as –The school goes to boy‖ is rejected by English syntactic analyzer.

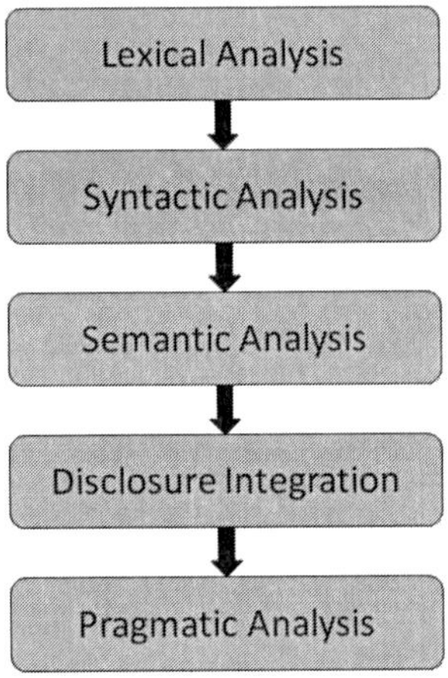

- **Semantic Analysis** – It draws the exact meaning or the dictionary meaning from the text. The text is checked for meaningfulness. It is done by mapping syntactic structures and objects in the task domain. The semantic analyzer disregards sentence such as –hot ice-cream‖.
- **Discourse Integration** – The meaning of any sentence depends upon the meaning of the sentence just before it. In addition, it also brings about the meaning of immediately succeeding sentence.
- **Pragmatic Analysis** – During this, what was said is reinterpreted on what it actually meant. It involves deriving those aspects of language which require real world knowledge.

Implementation Aspects of Syntactic Analysis

There are a number of algorithms researchers have developed for syntactic analysis, but we consider only the following simple methods –

- Context-Free Grammar

- Top-Down Parser

Let us see them in detail –

Context-Free Grammar

It is the grammar that consists rules with a single symbol on the left-hand side of the rewrite rules. Let us create grammar to parse a sentence –

–The bird pecks the grains‖

Articles (DET) – a | an | the

Nouns – bird | birds | grain | grains

Noun Phrase (NP) – Article + Noun | Article + Adjective

+ Noun

= DET N | DET ADJ N

Verbs – pecks | pecking | pecked

Verb Phrase (VP) – NP V | V NP

Adjectives (ADJ) – beautiful | small | chirping

The parse tree breaks down the sentence into structured parts so that the computer can easily understand and process it. In order for the parsing algorithm to construct this parse tree, a set of rewrite rules, which describe what tree structures are legal, need to be constructed.

These rules say that a certain symbol may be expanded in the tree by a sequence of other symbols. According to first order logic rule, if there are two strings Noun Phrase (NP) and Verb Phrase (VP), then the string combined by NP followed by VP is a sentence. The rewrite rules for the sentence are as follows –

S → NP VP

NP → DET N | DET ADJ N

VP → V NP Lexocon –

DET → a | the

ADJ → beautiful | perching

N → bird | birds | grain | grains

V → peck | pecks | pecking

The parse tree can be created as shown –

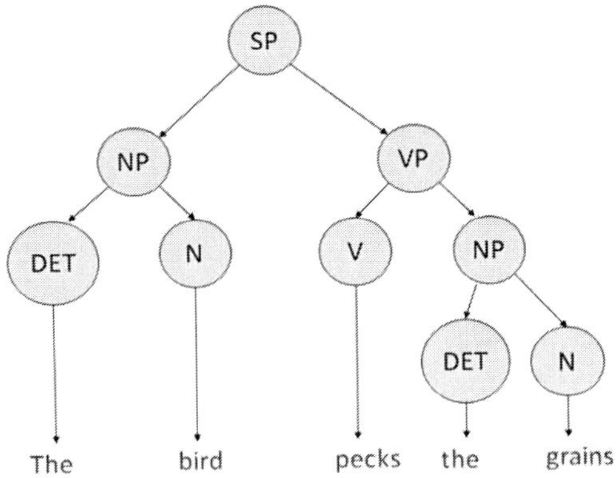

Now consider the above rewrite rules. Since V can be replaced by both, "peck" or "pecks", sentences such as "The bird peck the grains" can be wrongly permitted. i. e. the subject-verb agreement error is approved as correct.

Merit

The simplest style of grammar, therefore widely used one.

Demerits

- They are not highly precise. For example, –The grains peck the bird‖, is a syntactically correct according to parser, but even if it makes no sense, parser takes it as a correct sentence.
- To bring out high precision, multiple sets of grammar need to be prepared. It may require a completely different sets of rules for parsing singular and plural variations, passive sentences, etc., which can lead to creation of huge set of rules that are unmanageable.

Top-Down Parser

Here, the parser starts with the S symbol and attempts to rewrite it into a sequence of *terminal symbols* that matches the classes of the words in the input sentence until it consists entirely of terminal symbols.

These are then checked with the input sentence to see if it matched. If not, the process is started over again with a different set of rules. This is repeated until a specific rule is found which describes the structure of the sentence.

Merit

It is simple to implement.

Demerits

- It is inefficient, as the search process has to be repeated if an error occurs.
- Slow speed of working.

5.5 Machine Translation

Globalization and technology have revolutionized the international workforce. While most of the outcomes are positive, there are a few quirks that still need solving for. One thing that needs to be resolved is language processing and translation. These methods have improved throughout the last seventy years. Yet, there seems to always be more that technology and AI can do to perfect the machine translation process. When it comes to international business, a constant struggle is being able to effectively translate documents in a timely manner.

Globalization and technology have revolutionized the international workforce. While most of the outcomes are positive, there are a few quirks that still need solving for. One thing that needs to be resolved is language processing and translation. These methods have improved throughout the last seventy years. Yet, there seems to always be more that technology and AI can do to perfect the machine translation process. When it comes to international business, a constant struggle is being able to effectively translate documents in a timely manner. Technology has continued to grow and solve for this major pain point. While state of the art translation technology has been produced, it still remains to be the widely adopted. To understand the importance of this technology, it's first imperative that you understand the history of machine translation.

When human's translate documents it can take a very long time, which is why this isn't an ideal method for mass quantities. If you have high volume translation need using a machine is your best option. This is why in the 1950's machine translation techniques were initially developed. Machine translation is exactly what it sounds like. It's using a computer to translate documents with no human interaction. Did you know that automatic translation of texts from language to language was actually one of the first objectives for computers? Well, it was! Today it still remains to be a huge priority with the advancement of technology and AI. This technology was revolutionary at the time. Still, these approaches implemented by machine translation haven't always been entirely effective.

Machine translation has continued to challenge the world of technology, as time goes on we've adapted the ways in which we attempt to have AI fluently translate a variety of languages. At first, machine translation worked off of a –rule- based‖ system. This implemented a mixture of grammar and language rules as the basis of translation. In the 1990's translation began to use statistical methods, which generated translation by having computers analyze data from language use.

This development was imperative in creating a more effective way of machine translation. Still, it required prior information and data for AI to analyze. It, therefore, needed a translation alignment tool to be fully functional. Translate alignment by definition is the way in which technology can utilize already existing translation materials. This tool can syndicate your formerly translated documents and text into translation units which were a part of the data that statistical methods could pull from once this information was added into a translation memory.

While this was a step towards greater efficiency, machine translation and it's AI still depended on other variables to function accurately. Today, deep learning has provided technology that allows platforms to operate on their own. The most recent ways to translate have been developed by a deep neural network model which uses start-of-the-art technology to achieve incredibly accurate results.

Machine learning is a subdivision of AI and it operates in computer science. It typically pulls from statistical techniques to give computers (or machines) the ability to ―learn‖ and adapt their functions with data that is provided. This means that once the machine learns something, the AI can continue to progress, and function without specifically being programmed–or instructed–to do so. Deep learning is a component of machine learning methods established on learning through data, as explained above. In machine learning, this method has produced incredible results that allow technology to translate items more accurately than ever seen before.

Machine translation is an approach can be compared to Google Translate method. In the same way that Google Translate would, machine translation uses the statistical alignments of words and phrases to translate the text. This method can make sentences sound choppy and some words seem out of context, as machine translation attention doesn't mimic human behavior. When using technology to translate for you, you want it to sound like it's been done by a person and not artificial intelligence.

Humans don't use a statistical alignment approach to translate languages. Instead of translating by fragment, we take the time to fully understand the meaning of the phrases and words we are translating. The more technology grows the more machine translation alignment tools have become outdated.

Deep learning, a branch of machine learning methods, has transformed the way in which technology can efficiently translate documents. Neural or Deep Learning Machine Translation involves artificial intelligence and is a much more efficient method of translating. Their developments in machine language translation have made it easier for technology to translate from language to language.

Deep learning and Neural Machine Translation have developed a translation alignment tool based on the architecture called ―sequence to sequence‖. This state of the art technology mimics the human method of translating. In its processes it produces results that often time have much better quality, tightening the precision gap between human and machine translation. Making translations sound more fluid, natural, and most importantly accurate.

When these two methods are compared, you can see that the main difference is context. Machine translation takes a word and translates it with no regard for how it fits into the larger sentence. Neural Translation, however, looks at a text and analyses it for context. This new technology and AI allow for superior reading of the text prior to its translation. This function results in more accurate and fluent sounding results.

Each language has certain sayings that will not translate directly. For example in American English, a common phrase is –That was a piece of cake‖. When used, no one is actually talking about cake. Instead, it means the situation was handled easily. Colloquial sayings such as this are an example of why literal and word by word translation aren't always productive or accurate.

Now more than ever deep learning is needed. If you are in the international space and translating between languages is a priority for you, you should be utilizing this state of the art technology. The deep learning application of neural machine translation has created a machine translation alignment tool that can surpass all others.

It's predicted that only two years neural translation technology will eclipse the past two decades of translation materials. It already projects 50% fewer sentence arrangement mistakes, and 19% fewer grammatical mistakes. It's even started to specify gender in a different language–even though no one programmed this AI to do so.

A great example of this technology being utilized is by companies like Ureed, who aid in the processes of translation. They have a lot to gain from these new translation management systems and so do their clientele. Ureed has over 1 mil segments and 30+ million words available for translation that could train and develop deep learning model. This could massively enhance their already proficient editorial services by employing automatic machine translation to cut the time taken by human translators and editors and enhance the process. Which is why Ureed has partnered with Leading Point to use their state of the art technology to implement machine learning translation throughout its platform. Leading Point is leveraging the latest and proven machine learning technologies available in the market to further enhance the translation and editorial service provided by Ureed on its marketplace.

Ureed's internal development team can now implement this incredible technology to boost their software proficiency and accuracy when translating, reading and editing documents. The effectiveness of this machine translation deep learning technology provides an incredible opportunity for companies like Ureed who work in just an international and diverse climate. Language shouldn't have to be a barrier when it comes to business.

5.6 Speech Recognition

Speech is the most basic means of adult human communication. The basic goal of speech processing is to provide an interaction between a human and a machine. Speech processing system has mainly three tasks −

- **First**, speech recognition that allows the machine to catch the words, phrases and sentences we speak
- **Second**, natural language processing to allow the machine to understand what we speak, and
- **Third**, speech synthesis to allow the machine to speak.
- This chapter focuses on **speech recognition**, the process of understanding the words that are spoken by human beings. Remember that the speech signals are captured with the help of a microphone and then it has to be understood by the system.

Building a Speech Recognizer

Speech Recognition or Automatic Speech Recognition (ASR) is the center of attention for AI projects like robotics. Without ASR, it is not possible to imagine a cognitive robot interacting with a human. However, it is not quite easy to build a speech recognizer.

Difficulties in developing a speech recognition system Developing a high quality speech recognition system is really a difficult problem. The difficulty of speech recognition technology can be broadly characterized along a number of dimensions as discussed below –

- **Size of the vocabulary** – Size of the vocabulary impacts the ease of developing an ASR. Consider the following sizes of vocabulary for a better understanding.
    a. A small size vocabulary consists of 2-100 words, for example, as in a voice-menu system
    b. A medium size vocabulary consists of several 100s to 1,000s of words, for example, as in a database-retrieval task
    c. A large size vocabulary consists of several 10,000s of words, as in a general dictation task.
    c. Note that, the larger the size of vocabulary, the harder it is to perform recognition.
- **Channel characteristics** – Channel quality is also an important dimension. For example, human speech contains high bandwidth with full frequency range, while a telephone speech consists of low bandwidth with limited frequency range. Note that it is harder in the latter.
- **Speaking mode** – Ease of developing an ASR also depends on the speaking mode, that is whether the speech is in isolated word mode, or connected word mode, or in a continuous speech mode. Note that a continuous speech is harder to recognize.
- **Speaking style** – A read speech may be in a formal style, or spontaneous and conversational with casual style. The latter is harder to recognize.
- **Speaker dependency** – Speech can be speaker dependent, speaker adaptive, or speaker independent. A speaker independent is the hardest to build.
- **Type of noise** – Noise is another factor to consider while developing an ASR. Signal to noise ratio may be in

various ranges, depending on the acoustic environment that observes less versus more background noise –

- If the signal to noise ratio is greater than 30dB, it is considered as high range
- If the signal to noise ratio lies between 30dB to 10db, it is considered as medium SNR
- If the signal to noise ratio is lesser than 10dB, it is considered as low range

For example, the type of background noise such as stationary, non-human noise, background speech and crosstalk by other speakers also contributes to the difficulty of the problem.

- **Microphone characteristics** – The quality of microphone may be good, average, or below average. Also, the distance between mouth and micro-phone can vary. These factors also should be considered for recognition systems.
- Despite these difficulties, researchers worked a lot on various aspects of speech such as understanding the speech signal, the speaker, and identifying the accents. You will have to follow the steps given below to build a speech recognizer – Visualizing Audio Signals - Reading from a File and Working on it

This is the first step in building speech recognition system as it gives an understanding of how an audio signal is structured. Some common steps that can be followed to work with audio signals are as follows –

Recording

When you have to read the audio signal from a file, then record it using a microphone, at first.

Sampling

When recording with microphone, the signals are stored in a digitized form. But to work upon it, the machine needs them in the discrete numeric form. Hence, we should perform sampling at a certain frequency and convert the signal into the discrete numerical form. Choosing the high frequency for sampling implies that when humans listen to the signal, they feel it as a continuous audio signal.

Example

The following example shows a stepwise approach to analyze an audio signal, using Python, which is stored in a file. The frequency of this audio signal is 44,100 HZ.

Import the necessary packages as shown here –

```
import numpy as np
import matplotlib.pyplot as plt from scipy.io import wavfile
```

Now, read the stored audio file. It will return two values: the sampling frequency and the audio signal. Provide the path of the audio file where it is stored, as shown here –

```
frequency_sampling, audio_signal =
wavfile.read("/Users/admin/audio_file.wav")
```

Display the parameters like sampling frequency of the audio signal, data type of signal and its duration, using the commands shown –

```
print('\nSignal shape:', audio_signal.shape) print('Signal
Datatype:', audio_signal.dtype) print('Signal duration:',
round(audio_signal.shape[0] / float(frequency_sampling), 2),
'seconds')
```

This step involves normalizing the signal as shown below –

```
audio_signal = audio_signal / np.power(2, 15)
```

In this step, we are extracting the first 100 values from this signal to visualize. Use the following commands for this purpose –

```
audio_signal = audio_signal [:100]
time_axis = 1000 * np.arange(0, len(signal), 1) /
float(frequency_sampling)
Now, visualize the signal using the commands given below –
```

```
plt.plot(time_axis, signal, color='blue') plt.xlabel('Time
(milliseconds)') plt.ylabel('Amplitude')
plt.title('Input audio signal')
plt.show()
```

You would be able to see an output graph and data extracted for the above audio signal as shown in the image here

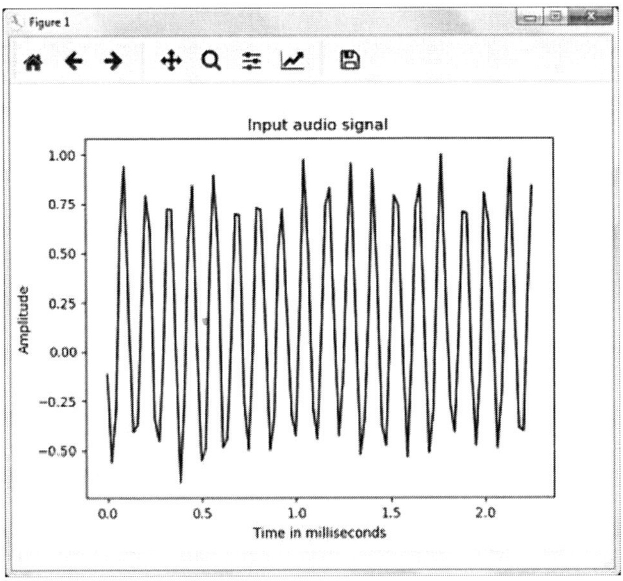


```
Signal shape: (132300,) Signal Datatype: int16
Signal duration: 3.0 seconds
```

Characterizing the Audio Signal: Transforming to Frequency Domain Characterizing an audio signal involves converting the time domain signal into frequency domain, and understanding its frequency components, by. This is an important step because it gives a lot of information about the signal. You can use a mathematical tool like Fourier Transform to perform this transformation.

Example

The following example shows, step-by-step, how to characterize the signal, using Python, which is stored in a file. Note that here we are using Fourier Transform mathematical tool to convert it into frequency domain.

Import the necessary packages, as shown here –

```
import numpy as np
import matplotlib.pyplot as plt from scipy.io import wavfile
```

Now, read the stored audio file. It will return two values: the sampling frequency and the the audio signal. Provide the path of the audio file where it is stored as shown in the command here –

```
frequency_sampling, audio_signal =
wavfile.read("/Users/admin/sample.wav")
```

In this step, we will display the parameters like sampling frequency of the audio signal, data type of signal and its duration, using the commands given below –

```
print('\nSignal shape:', audio_signal.shape) print('Signal
Datatype:', audio_signal.dtype) print('Signal duration:',
round(audio_signal.shape[0] /
float(frequency_sampling), 2), 'seconds')
```

In this step, we need to normalize the signal, as shown in the following command –

```
audio_signal = audio_signal / np.power(2, 15)
```

This step involves extracting the length and half length of the signal. Use the following commands for this purpose –

```
length_signal = len(audio_signal)


half _length = np.ceil((length_signal + 1) /
2.0).astype(np.int)
```

Now, we need to apply mathematics tools for transforming into frequency domain. Here we are using the Fourier Transform.

```
signal_frequency = np.fft.fft(audio_signal)
```

Now, do the normalization of frequency domain signal and square it –

```
signal_frequency =
abs(signal_frequency[0:half_length]) / length_signal
signal_frequency **= 2
```

Next, extract the length and half length of the frequency transformed signal –

```
len_fts = len(signal_frequency)
```

Note that the Fourier transformed signal must be adjusted for even as well as odd case.

```
if length_signal % 2: signal_frequency[1:len_fts] *= 2 else:
signal_frequency[1:len_fts-1] *= 2
```

Now, extract the power in decibal(dB) –

```
signal_power = 10 * np.log10(signal_frequency)
```

Adjust the frequency in kHz for X-axis –

```
x_axis = np.arange(0, len_half, 1) * (frequency_sampling/
length_signal) / 1000.0
```

Now, visualize the characterization of signal as follows –

```
plt.figure()
plt.plot(x_axis, signal_power, color='black')
plt.xlabel('Frequency (kHz)') plt.ylabel('Signal power (dB)')
plt.show()
```

You can observe the output graph of the above code as shown in the image below –

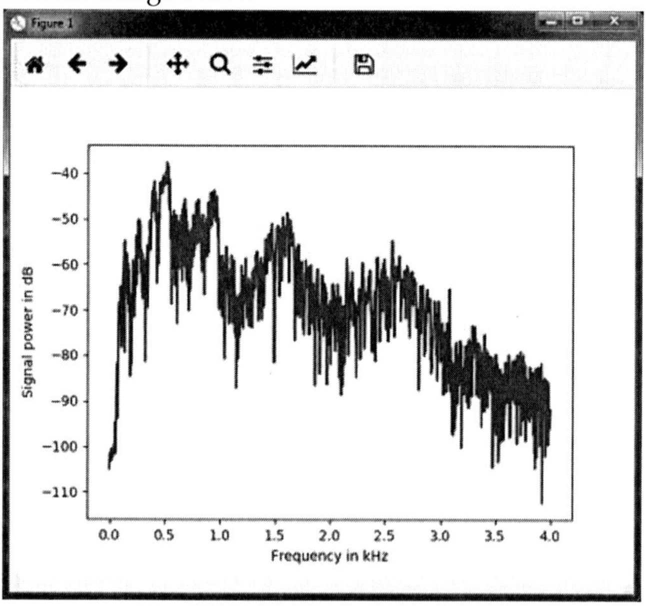

Generating Monotone Audio Signal

The two steps that you have seen till now are important to learn about signals. Now, this step will be useful if you want to generate the audio signal with some predefined parameters. Note that this step will save the audio signal in an output file.

Example

In the following example, we are going to generate a monotone signal, using Python, which will be stored in a file. For this, you will have to take the following steps –

Import the necessary packages as shown -

```
import numpy as np
import matplotlib.pyplot as plt from scipy.io.wavfile import
write
```

Provide the file where the output file should be saved
output_file = 'audio_signal_generated.wav'

Now, specify the parameters of your choice, as shown -

```
duration = 4 # in seconds frequency_sampling = 44100 # in Hz
frequency_tone = 784
min_val = -4 * np.pi
max_val = 4 * np.pi
```

In this step, we can generate the audio signal, as shown -

```
t = np.linspace(min_val, max_val, duration *
frequency_sampling)
audio_signal = np.sin(2 * np.pi * tone_freq * t)
```

Now, save the audio file in the output file – write(output_file, frequency_sampling, signal_scaled) Extract the first 100 values for our graph, as shown –

```
audio_signal = audio_signal[:100]
time_axis = 1000 * np.arange(0, len(signal), 1) /
float(sampling_freq)
```

Now, visualize the generated audio signal as follows -

```
plt.plot(time_axis, signal, color='blue') plt.xlabel('Time in
milliseconds') plt.ylabel('Amplitude') plt.title('Generated
audio signal')
plt.show()
```

You can observe the plot as shown in the figure given here -

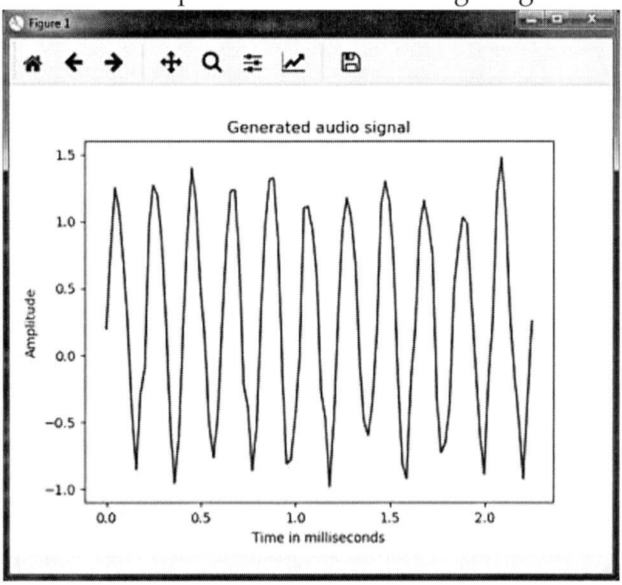

Feature Extraction from Speech

This is the most important step in building a speech recognizer because after converting the speech signal into the frequency domain, we must convert it into the usable form of feature vector. We can use different feature extraction techniques like MFCC, PLP, PLP-RASTA etc. for this purpose.

Example

In the following example, we are going to extract the features from signal, step-by-step, using Python, by using MFCC technique.

Import the necessary packages, as shown here –

```
import numpy as np
import matplotlib.pyplot as plt from scipy.io import wavfile
from python_speech_features import mfcc, logfbank
```

Now, read the stored audio file. It will return two values – the sampling frequency and the audio signal. Provide the path of the audio file where it is stored.

```
frequency_sampling, audio_signal =
wavfile.read("/Users/admin/audio_file.wav")
```

Note that here we are taking first 15000 samples for analysis.

```
audio_signal = audio_signal[:15000]
```

Use the MFCC techniques and execute the following command to extract the MFCC features –

features_mfcc = mfcc(audio_signal, frequency_sampling)

Now, print the MFCC parameters, as shown –

```
print('\nMFCC:\nNumber of windows =',
features_mfcc.shape[0])
print('Length of each feature =', features_mfcc.shape[1])
```

Now, plot and visualize the MFCC features using the commands given below –

```
features_mfcc = features_mfcc.T
plt.matshow(features_mfcc)
plt.title('MFCC')
```

In this step, we work with the filter bank features as shown Extract the filter bank features –

```
filterbank_features = logfbank(audio_signal,
frequency_sampling)
```

Now, print the filterbank parameters.

```
print('\nFilter bank:\nNumber of windows =',
filterbank_features.shape[0])
print('Length of each feature =',
filterbank_features.shape[1])
```

Now, plot and visualize the filterbank features.

```
filterbank_features = filterbank_features.T
plt.matshow(filterbank_features) plt.title('Filter bank')
plt.show()
```

As a result of the steps above, you can observe the following outputs: Figure1 for MFCC and Figure2 for Filter Bank

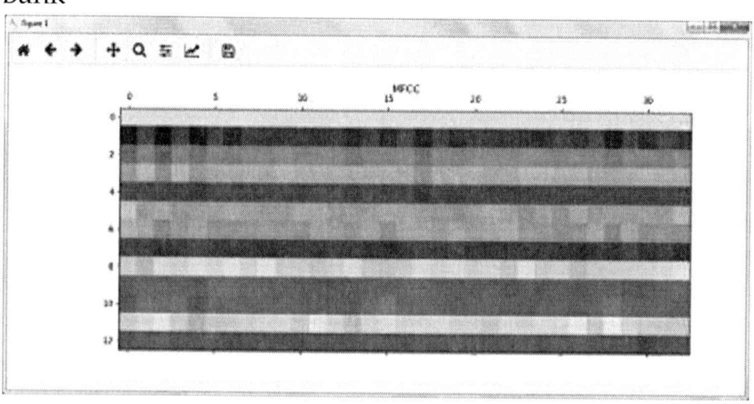

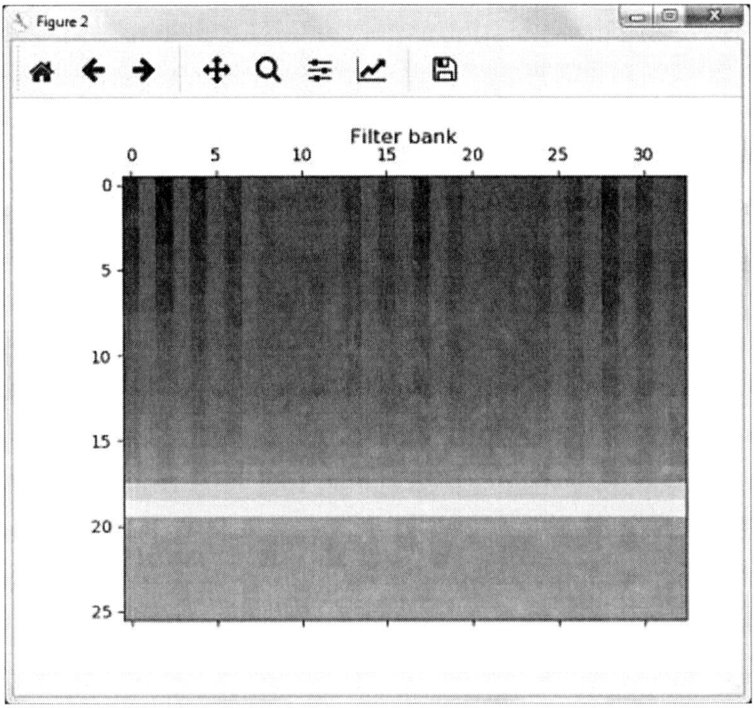

Recognition of Spoken Words

Speech recognition means that when humans are speaking, a machine understands it. Here we are using Google Speech API in Python to make it happen. We need to install the following packages for this –

- **Pyaudio** – It can be installed by using **pip install Pyaudio** command.
- **Speech Recognition** – This package can be installed by using **pip install Speech Recognition.**
- **Google-Speech-API** – It can be installed by using the command **pip install google-api-python-client**.

Example

Observe the following example to understand about recognition of spoken words –

Import the necessary packages as shown –

```
import speech_recognition as sr
```

Create an object as shown below -

```
recording = sr.Recognizer()
```

Now, the **Microphone()** module will take the voice as input -

```
with sr.Microphone() as source:
recording.adjust_for_ambient_noise(source) print("Please Say something:")
audio = recording.listen(source)
```

Now google API would recognize the voice and gives the output.

5.7 Robot

Robotics is a branch of AI, which is composed of Electrical Engineering, Mechanical Engineering, and Computer Science for designing, construction, and application of robots.

Aspects of Robotics

- The robots have **mechanical construction**, form, or shape designed to accomplish a particular task.
- They have **electrical components** which power and control the machinery.
- They contain some level of **computer program** that determines what, when and how a robot does something.

Difference in Robot System and Other AI Program

Here is the difference between the two –

AI Programs	Robots
They usually operate in computer-stimulated worlds.	They operate in real physical world
The input to an AI program is in symbols and rules.	Inputs to robots is analog signal in the form of speech waveform or images
They need general purpose computers to operate on.	They need special hardware with sensors and effectors.

Robot Locomotion

Locomotion is the mechanism that makes a robot capable of moving in its environment. There are various types of locomotions –

- Legged
- Wheeled
- Combination of Legged and Wheeled Locomotion
- Tracked slip/skid

Legged Locomotion

- This type of locomotion consumes more power while demonstrating walk, jump, trot, hop, climb up or down, etc.
- It requires more number of motors to accomplish a movement. It is suited for rough as well as smooth terrain where irregular or too smooth surface makes it consume more power for a wheeled locomotion. It is little difficult to implement because of stability issues.
- It comes with the variety of one, two, four, and six legs. If a robot has multiple legs then leg coordination is necessary for locomotion.

The total number of possible **gaits** (a periodic sequence of lift and release events for each of the total legs) a robot can travel depends upon the number of its legs.

If a robot has k legs, then the number of possible events N = (2k-1)!.

In case of a two-legged robot (k=2), the number of possible events is N = (2k-1)! = (2*2-1)! = 3! = 6.
Hence there are six possible different events –

- Lifting the Left leg
- Releasing the Left leg
- Lifting the Right leg
- Releasing the Right leg
- Lifting both the legs together
- Releasing both the legs together

In case of k=6 legs, there are 39916800 possible events. Hence the complexity of robots is directly proportional to the number of legs.

Wheeled Locomotion

It requires fewer number of motors to accomplish a movement. It is little easy to implement as there are less stability issues in case of more number of wheels. It is power efficient as compared to legged locomotion.

- **Standard wheel** – Rotates around the wheel axle and
- around the contact

- **Castor wheel** – Rotates around the wheel axle and the offset steering joint.
- **Swedish 45° and Swedish 90° wheels** – Omni-wheel, rotates around the contact point, around the wheel axle, and around the rollers.
- **Ball or spherical wheel** – Omnidirectional wheel, technically difficult to implement.

Slip/Skid Locomotion

In this type, the vehicles use tracks as in a tank. The robot is steered by moving the tracks with different speeds in the same or opposite direction. It offers stability because of large contact area of track and ground.

Components of a Robot

Robots are constructed with the following –

- **Power Supply** – The robots are powered by batteries, solar power, hydraulic, or pneumatic power sources.

- **Actuators** – They convert energy into movement.
- **Electric motors (AC/DC)** – They are required for

rotational movement.

- **Pneumatic Air Muscles** – They contract almost 40% when air is sucked in them.
- **Muscle Wires** – They contract by 5% when electric current is passed through them.
- **Piezo Motors and Ultrasonic Motors** – Best for industrial robots.
- **Sensors** – They provide knowledge of real time information on the task environment. Robots are equipped with vision sensors to be to compute the depth in the environment. A tactile sensor imitates the mechanical properties of touch receptors of human fingertips.

Computer Vision

This is a technology of AI with which the robots can see. The computer vision plays vital role in the domains of safety, security, health, access, and entertainment.

Computer vision automatically extracts, analyzes, and comprehends useful information from a single image or an array of images. This process involves development of algorithms to accomplish automatic visual comprehension.

Hardware of Computer Vision System

This involves –

- Power supply
- Image acquisition device such as camera
- A processor
- A software
- A display device for monitoring the system
- Accessories such as camera stands, cables, and connectors

Tasks of Computer Vision

- **OCR** – In the domain of computers, Optical Character Reader, a software to convert scanned documents into editable text, which accompanies a scanner.
- **Face Detection** – Many state-of-the-art cameras come with this feature, which enables to read the face and take the picture of that perfect expression. It is used to let a user access the software on correct match.
- **Object Recognition** – They are installed in supermarkets, cameras, high-end cars such as BMW, GM, and Volvo.
- **Estimating Position** – It is estimating position of an object with respect to camera as in position of tumor in human's body.

Application Domains of Computer Vision

- Agriculture
- Autonomous vehicles
- Biometrics
- Character recognition
- Forensics, security, and surveillance
- Industrial quality inspection
- Face recognition
- Gesture analysis
- Geoscience
- Medical imagery
- Pollution monitoring
- Process control
- Remote sensing
- Robotics
- Transport

Applications of Robotics

The robotics has been instrumental in the various domains such as –

- **Industries** – Robots are used for handling material, cutting, welding, color coating, drilling, polishing, etc.
- **Military** – Autonomous robots can reach inaccessible and hazardous zones during war. A robot named *Daksh*, developed by Defense Research and Development Organization (DRDO), is in function to destroy life-threatening objects safely.
- **Medicine** – The robots are capable of carrying out hundreds of clinical tests simultaneously, rehabilitating permanently disabled people, and performing complex surgeries such as brain tumors.
- **Exploration** – The robot rock climbers used for space exploration, underwater drones used for ocean exploration are to name a few.
- **Entertainment** – Disney's engineers have created hundreds of robots for movie making.

5-QUESTION BANK
PART - A

1. Define planning.

Planning can be viewed as a type of problem solving in which the agent uses beliefs about actions and their consequences to search for a solution.

2. What are the features of an ideal planner?

   a. The planner should be able to represent the states, goals and actions.
   b. The planner should be able to add new actions at any time.
   c. The planner should be able to use Divide and Conquer method for solving very big problems.

3. What are the components that are needed for representing an action?

The components that are needed for representing an action are: Action description. Precondition. Effect.

4. What are the components that are needed for representing a plan?

The components that are needed for representing a plan are: A set of plans steps. A set of ordering constraints.

5. What are the different types of planning? The different types of planning are as follows:

   a. Situation space planning.
   b. Progressive planning. c. Regressive planning.
   c. Partial order planning.
   d. Fully instantiated planning.

6. What are the ways in which incomplete and incorrect information's can be handled in planning?

They can be handled with the help of two planning agents namely,

a. Conditional planning agent.
b. Replanning agent.

7. Define a solution.

A solution is defined as a plan that an agent can execute and that guarantees the achievement of goal.

8. Define a complete plan.

A complete plan is one in which every precondition of every step is achieved by some other step.

9. Define a consistent plan.

A consistent plan is one in which there are no contradictions in the ordering or binding constraints.

10. Define conditional planning.

Conditional planning is a way in which the incompleteness of information is incorporated in terms of adding a conditional step, which involves if – then rules.

11. Give the classification of learning process. The learning process can be classified as:

Process which is based on coupling new information to previously acquired knowledge.

a. Learning by correcting mistakes.
b. Learning by explaining experience. c. Learning by analyzing differences d. Learning by managing models

12. State the wait and see principle.

The law states that, -When there is doubt about what to do, do nothing‖

13. State the no altering principle.

The law states that, - When an object or situation known to be an example, fails to match a general model, create a special case exception model‖.

14. State Martin's law.

The law states that, –You cannot learn anything unless you almost know it already‖.

15. Define Similarity nets.

Similarity net is an approach for arranging models. Similarity net is a representation in which nodes denotes models, links connect similar models and links are tied to different descriptions.

16. Define Reification.

The process of treating something abstract and difficult to talk about as though it were concrete and easy to talk about is called as reification.

17. What is reified link?

The elevation of a link to the status of a describable node is a kind of reification. When a link is so elevated then it is said to be a reified link..

18. Define Communication.

Communication is the international exchange of information brought about by the production and perception of signs drawn from a shared system of conventional signs.

19. Define Language.

Language enables us to communicate most of what we have observed about the environment.

20. Define Formal Language.

A formal language is defined as a set of strings of terminal symbols. It is called as words.